Beginning the New Geography

Ken Briggs

Hodder and Stoughton
London Sydney Auckland Toronto

IMPORTANT NOTE

Some of the exercises in this book require the use of 'blank' diagrams. These can be obtained by making photocopies – free of normal copyright restrictions – from the set of 10 cards (*Beginning the New Geography: Cards* ISBN 0 340 23311 7), obtainable from the publishers through your usual supplier.

British Library Cataloguing in Publication Data

Briggs, Kenneth
Beginning the new geography. – (Introducing the new geography).
1. Geography – Text-books – 1945–
I. Title II. Series
910′.01 G128

ISBN 0 340 21926 2

First published 1979

Illustrations by John Baxter and Vic Brand

Printed in Great Britain for
Hodder and Stoughton Educational,
a division of Hodder and Stoughton Ltd,
Mill Road, Dunton Green, Sevenoaks, Kent,
by Fletcher & Son Ltd, Norwich

Preface

It appears that the existing books in the 'Introducing the New Geography' series have made a considerable impact upon the development of modern geography in schools in the lower school and even in the sixth form. But various reviewers have commented that these books demand a considerable amount of application on the part of pupils.

This book is designed as a relatively simple introductory course on locational analysis in geography which will function as an ideal introduction to these more specialised books and which can be used profitably by a wide range of pupils of various levels of ability. The book consists of a series of practical exercises, none of which necessitate writing or drawing in the book itself, which together make up an introductory course in modern geography. Modern concepts are built up in a logical order by means of a series of carefully graded exercises, but it is possible for the teacher to select Units and Exercises which are appropriate to his own particular syllabus. Alternatively, he could regard the book as the core of a year's work in geography and supplement it with additional material of his own.

Within each Section, and within the book as a whole, the Units become progressively more difficult by easy stages. The book will be of particular value when introducing essential geographical concepts to children who find difficulties in expressing themselves verbally.

K. BRIGGS

Contents

Section A
Points in Geography

Unit I: How points are arranged along lines

EXERCISE 1: Points on lines

Points or dots on a map can stand for many different things. We can mark a dot on a map or diagram to represent a school, a town, or even the earth itself if we only want to show *where* it is and not what shape it is. You will see the earth as a dot on diagrams of the whole Solar System. On motoring maps towns are commonly shown as dots. It all depends upon the *scale* of the map. If a map of the whole world were to be drawn on this page even a city as big as London would have to be shown as a dot.

Sometimes these dots are arranged in distinct lines. Examples are shops or houses along a street, intersections along a motorway, towns along a river or a coastline.

(i) Copy Figure 1 and draw the pattern which you think would be formed by a queue of 40 people waiting to enter the cinema. Let each dot represent one person.

(ii) On Figure 2 cars can only enter the supermarket car park where the arrows are shown. Copy the map and draw the pattern which you think would be formed by the first 40 cars to arrive in the morning. Let each dot represent one car. Each small rectangle can hold one car.

(iii) On Figure 3 three intersections allow cars to enter the motorway. At which three of the letters A to G do you think these intersections are located? Compare your choice with other members of your class.

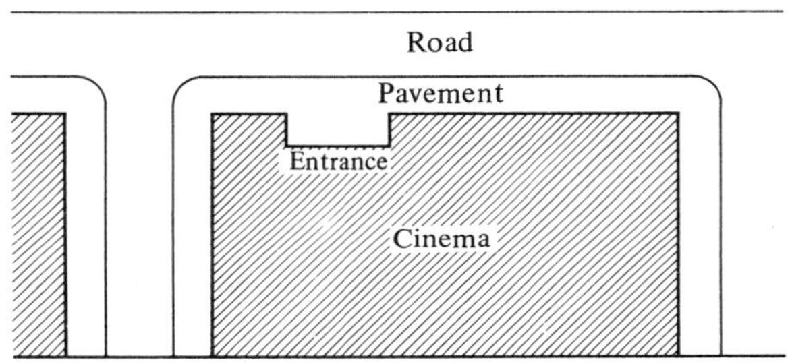

Figure 1

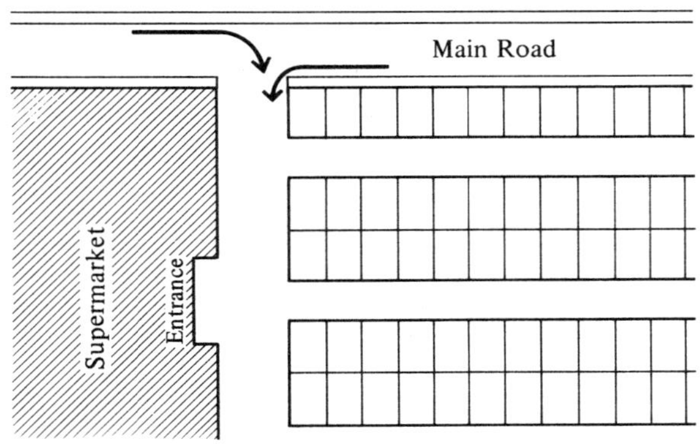

Figure 2

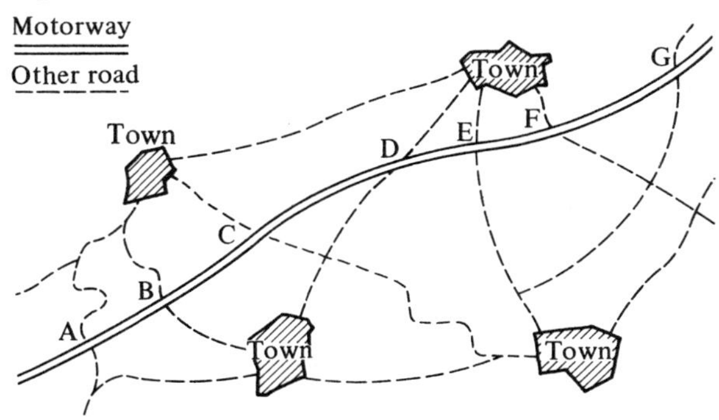

Figure 3

EXERCISE 2: Clustered and regular distributions

You will have noticed in the previous exercise that points arranged in lines can form very different patterns. Sometimes the points are very close to one another, as in the cinema queue and the car park. This is called a *clustered distribution.*

Figure 4 shows two different kinds of *clustered* distributions. In one case the dots form one big cluster, each dot being one kilometre away from the one nearest to it. In the other example, the dots form *two* clusters, but once again each dot is one kilometre away from its nearest neighbour. In both of these patterns the *average nearest neighbour distance* is one kilometre.

A different kind of distribution is one in which the dots are as far apart as they can possibly be. This is a *regular* distribution as shown in Figure 5. In this case each dot is three kilometres away from its nearest neighbour. The *average nearest neighbour distance* is therefore three kilometres. It is not possible to fit 10 dots on a line of this length so that their average nearest neighbour distance is more than three kilometres. Try an experiment on a piece of scrap paper.

Very strongly clustered and perfectly regular distributions are unlikely to exist unless they have been deliberately planned. Real-life distributions will usually be irregular, as in Figure 6.

Study Figure 6 and make a list of the distances from each dot to its nearest neighbour. The arrow near each dot shows the direction to its nearest neighbour. Then work out the average nearest neighbour distance of the dots. If your answer is 2·5 kilometres or over, the distribution can be described as 'regular' even though it is not perfectly regular.

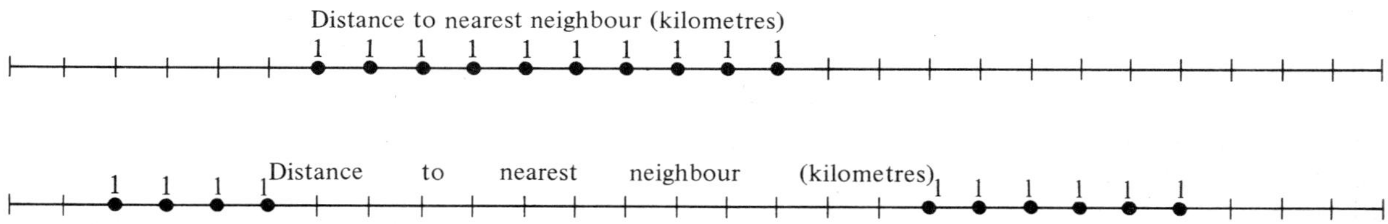

Examples of clustered distributions

Figure 4

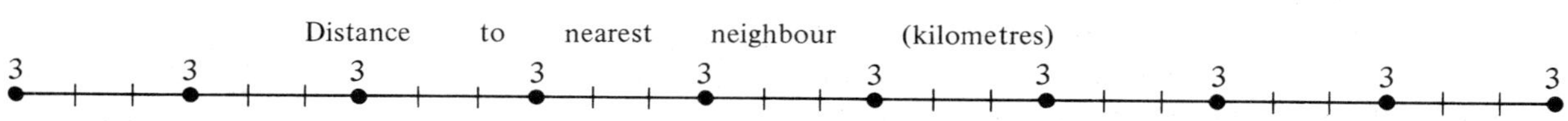

Example of a regular distribution

Figure 5

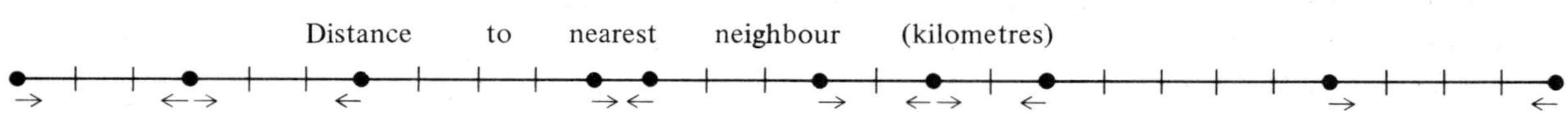

Figure 6

EXERCISE 3: Intersections along the M6 motorway

The map (Figure 7) shows the part of the M6 motorway between Preston in the south and Carlisle in the north. Between Preston, on the fringe of the industrial area of south Lancashire, and Kendal, the motorway passes through a fairly well populated agricultural lowland. Between Kendal and Penrith, it threads its way between high mountains in a very sparsely populated area. North of Penrith, it once again passes through an agricultural area of moderate population density.

(i) Which two intersections would you expect to be least used by traffic?

(ii) Considering all the information on the map, which intersection would you expect to be busiest on a fine Sunday morning in midsummer?

(iii) Let us now test whether these intersections as a whole form a clustered or a regular pattern.

Make a list of all the distances from each intersection to its nearest neighbour. Then work out the average nearest neighbour distance. If the average nearest neighbour distance is over 8·2 kilometres the distribution is *regular*. If it is under 3·1 kilometres, the distribution is *clustered*. If it is between 3·1 km and 8·2 km it is neither definitely clustered nor definitely regular, but is best described as *irregular*. How would you describe the distribution of intersections on this part of the M6?

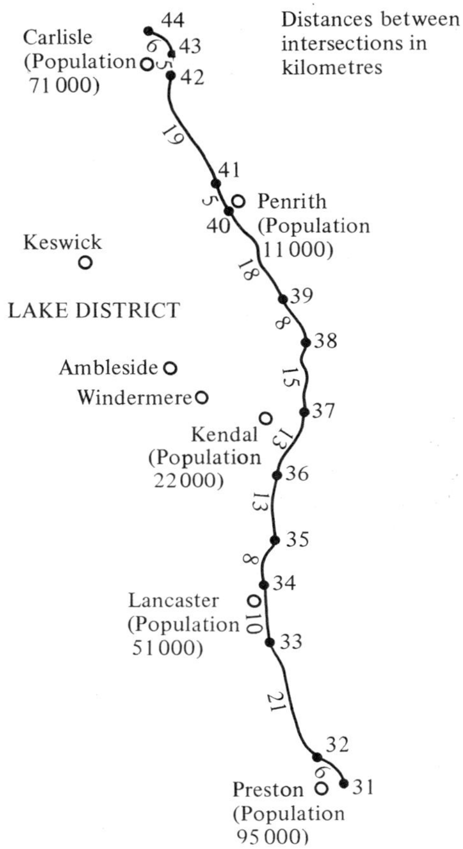

Figure 7

Unit II: Market areas (1)

EXERCISE 1: Shop location

Many shops, such as the local butcher or greengrocer, sell their goods mainly to the people who live round about. Such shops are usually scattered fairly evenly throughout a town, so that their customers do not have to travel very far to visit them. The area from which a shop draws its customers is called its *market area*.

Suppose there is a village consisting of a single long line of houses, as shown in Figure 8. A greengrocer wants to open a shop in such a place that his customers in the village are as near as possible to his shop. Clearly the best place for the shop is at the centre of the line of houses. The whole village would be the market area of the shop.

If he had wanted to open two shops he would not have placed them both in the centre of the

Figure 8

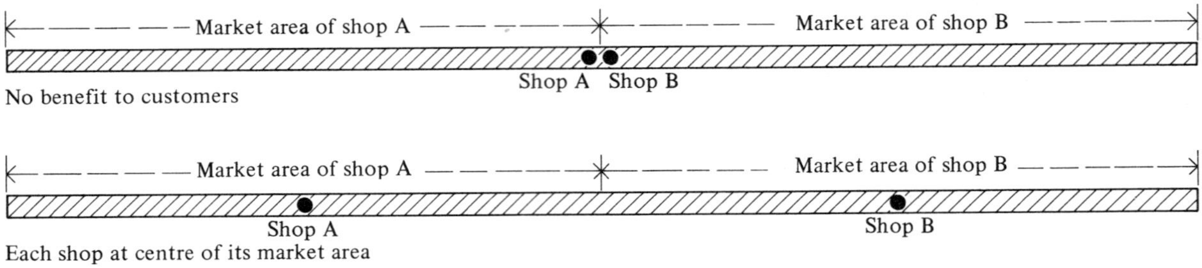

Figure 9

village. This would have been of no extra benefit to his customers. Everybody would have had to walk very nearly as far as with one shop. With two shops he should place them as shown in Figure 9, so that no customer would need to walk very far.

(i) Draw a diagram like these to show where he should set up three shops so as to save his customers from walking further than necessary. Label the market areas of the three shops.

(ii) Now draw similar diagrams for four shops, and for six shops.

EXERCISE 2: Competition between shops

Suppose now that two different greengrocers, Adam and Brown, decide to begin to open shops in a similar village. The villagers all decide that they will buy from the shop nearest to their home whether it belongs to Adam or Brown.

So Adam and Brown try to locate their shops in the village so that their market areas are as large as possible.
Adam opens his first shop (A1) at the centre of the village so as to be as near as possible to all the houses in the village (Fig. 10).

Brown then opens his first shop (B1) as shown in Figure 11.

Brown's choice of location for his shop does not seem a very good one. Perhaps you would have chosen a different place.

(i) Draw a diagram to show where you think Brown should have opened his first shop.

Now Adam opens his second shop (A2) as shown in Figure 12.

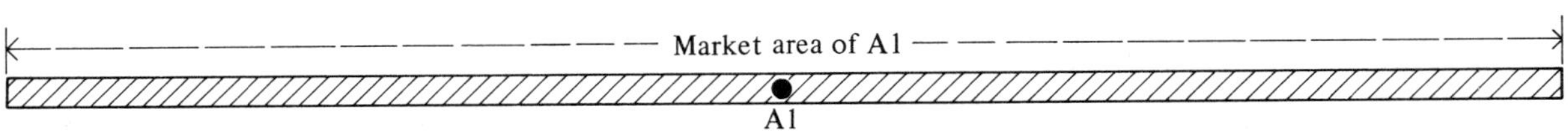

Figure 10

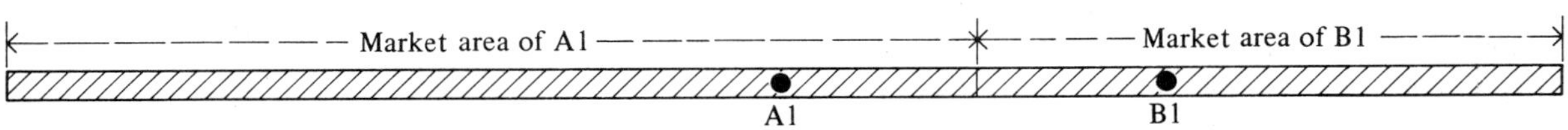

Figure 11

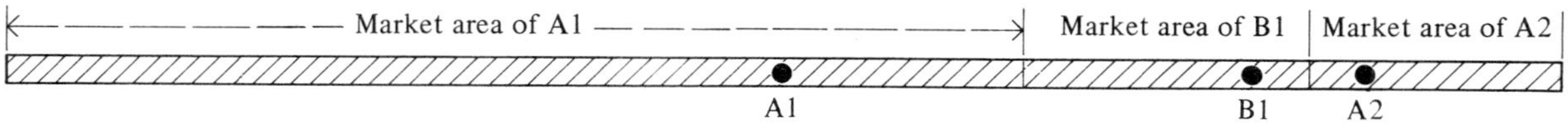

Figure 12

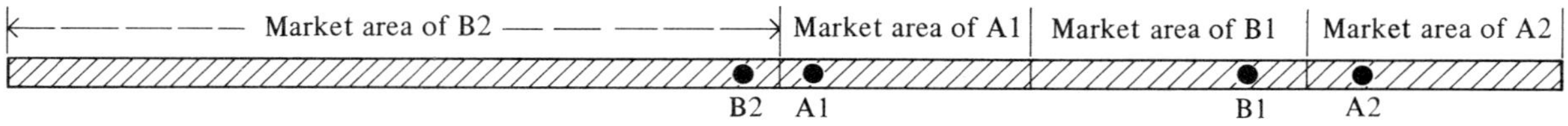

Figure 13

Brown opens his second shop (B2) as shown in Figure 13.

Brown has chosen an excellent place for his second shop (B2). It has been able to capture almost the whole of the market area of shop A1. Brown now is almost certain to be enjoying a greater amount of trade than Adam.

(ii) Draw a diagram to show where you think Adam's third shop (A3) should be. Mark and label all the market areas.

EXERCISE 3: Shop competition game (1)

In this game, as in the two previous exercises, we imagine that a village consists of a single long line of houses, but in this case the outline of each house is shown. All the villagers, as before, buy at the shop which is nearest to their home.

Each player in the game (A and B) sets up a shop in turn by occupying one of the houses. It is best to toss a coin to decide which player sets up the first shop. Then the players set up shops alternately until they each have three shops. The winner of the game is the player whose market areas for all his three shops contain the most houses.

Figure 14 shows the outcome of a game played by two players (A and B) showing the method of scoring. It can be seen that Player A wins by 13 houses to 12.

Try this game with your friends using blank maps of the village.

EXERCISE 4: Shop competition game (2)

Now try the same game but with the houses arranged in different patterns. You will find that in these cases it is even more important to place every one of your shops in the best possible position. When working out the score you must count *along* the rows of houses. You cannot short-cut across the spaces between.

From playing these games, give a short explanation of what you think is a good way to place your shops so that you have the best possible chance of winning.

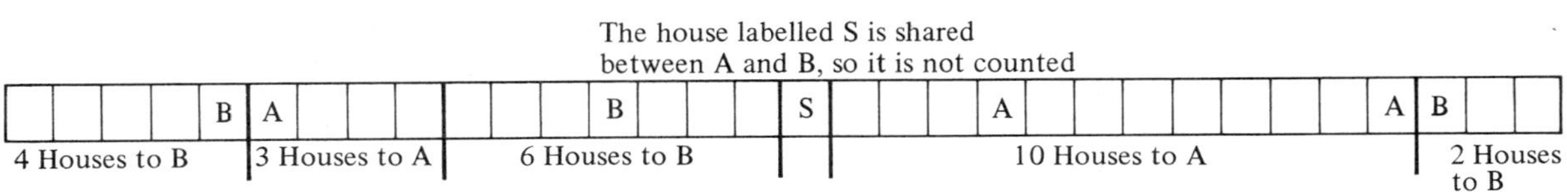

Figure 14

Unit III: How points are arranged in space

EXERCISE 1: Clustered and regular distributions

How would you set out the plants if you were planting a flower bed in spring? You would not want the plants to be too close together or they might starve one another of food and light, but you would want the bed to hold as many plants as possible so as to give a bright display of flowers.

Figures 15 to 17 show three flower beds of exactly the same size. Each dot is one plant. Let us compare these planting patterns.

(i) What are the disadvantages of the planting pattern shown in Figure 15?

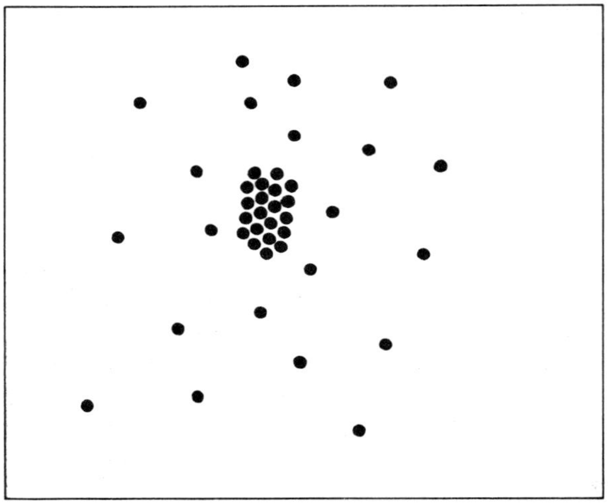

Clustered distribution 42 Plants

Figure 15

The planting patterns shown in Figures 16 and 17 are alike in one respect. That is that each plant is the same distance from its nearest neighbour. The plants in Figure 16 are arranged in a square pattern (look at the dotted square). The plants in Figure 17 are arranged in a triangular pattern, as shown by the dotted triangle.

(ii) Giving reasons, explain why the triangular pattern (Figure 17) is better than the square pattern (Figure 16).

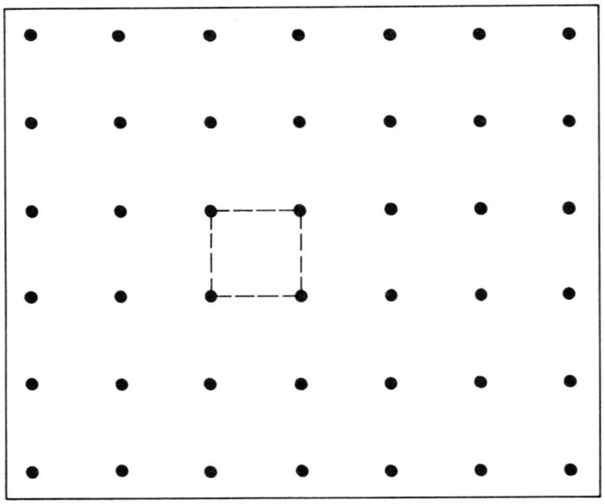

Regular distribution (square pattern) 42 Plants

Figure 16

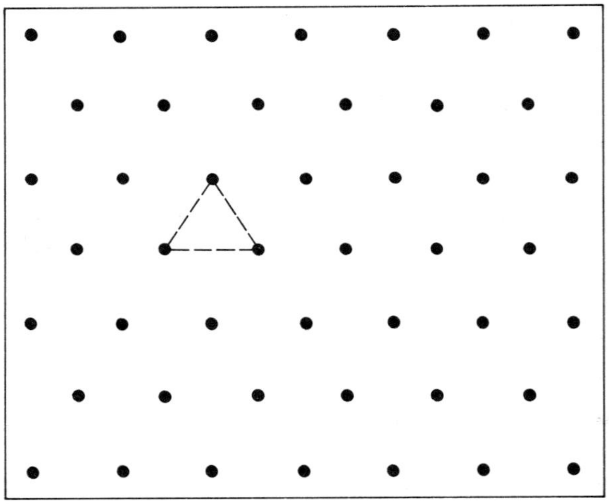

Regular distribution (triangular pattern) 46 Plants

Figure 17

(iii) Try to think of other real-life examples of dot patterns like Figures 15, 16 and 17.

EXERCISE 2: Irregular or chance distributions

Many geographical distributions cannot easily be classified as regular or clustered, like the one in the previous exercise. Many are somewhere between these two extremes and show some of the characteristics of each.

Let us now look at a distribution which might easily be the result of pure chance. A good example might be the pattern produced by the first few raindrops of a shower on a patch of pavement. It would certainly be an *irregular* pattern.

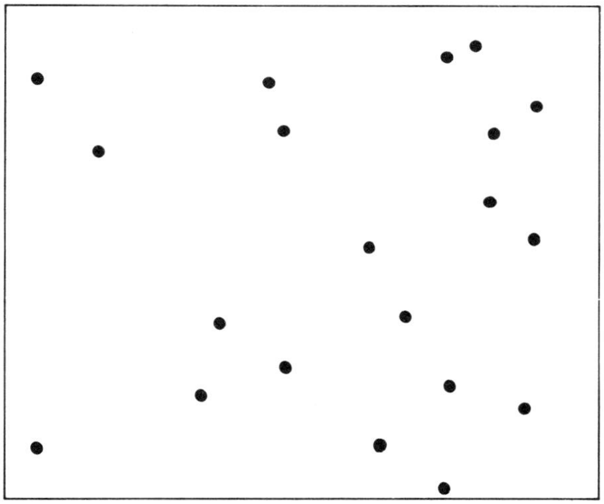

Irregular or chance distribution

Figure 18

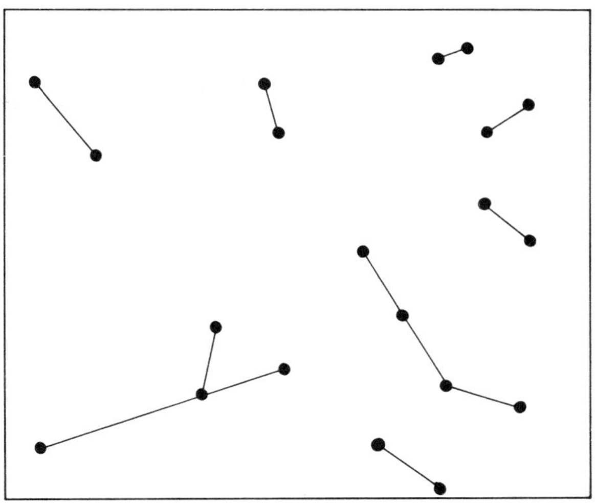

Nearest neighbours

Figure 19

A distribution of this sort is shown in Figure 18. You will notice that:

(i) the dots are fairly well scattered over the area shown,

(ii) some of the dots tend to form small clusters. These clusters are more easily recognized by drawing nearest neighbour lines, as in Figure 19.

Draw some diagrams with scattered dots, like Figure 18, and practise drawing nearest neighbour lines.

We can now summarize the main kinds of dot distributions as follows:

(i) *Regular distributions* (as in Figures 16 and 17)
It is almost impossible for these to have been produced by chance. There must be some *factor* which has caused them to be regular.

(ii) *Strongly clustered distributions* (as in Figure 15)
These again are very unlikely to have been produced by chance. It is much more likely that some *factor* has attracted the dots to the place where they are clustered.

(iii) *Irregular distributions* (as in Figure 18)
There may be a strong probability that these could be the result of pure chance.

But of course each of these distribution types merges gradually into another. There can be a gradual change from 'perfectly clustered' to 'perfectly regular'. In studying dot distributions we have to decide whether a pattern is sufficiently clustered or regular to make it very unlikely that it could have been produced by chance.

EXERCISE 3: More kinds of distributions

Now construct some dot distributions for yourself.

(i) Copy the diagrams on this page and mark 22 dots on each as follows:

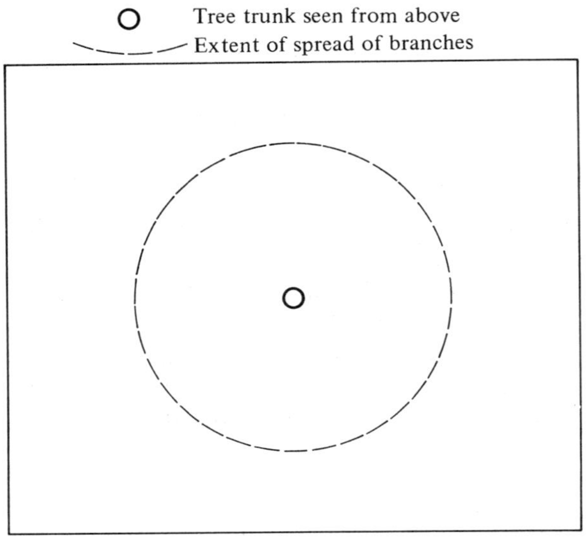

Figure 20

Mark 22 leaves which have fallen from the tree on a windless autumn day (Fig. 20).

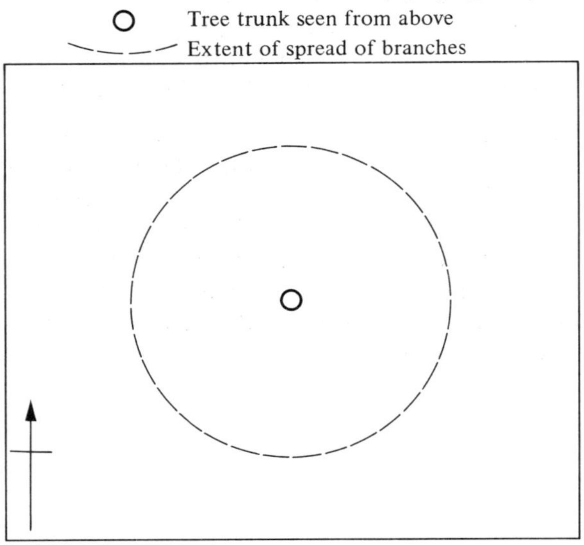

Figure 21

Mark 22 leaves which have fallen from the tree when a west wind is blowing (Fig. 21).

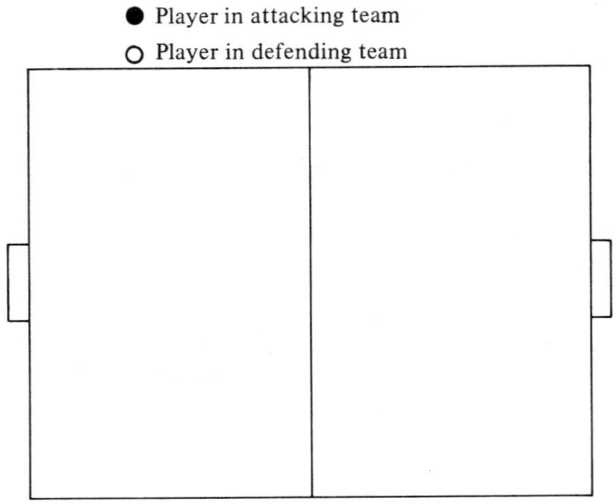

Figure 22

Mark 22 players taking part in a football or hockey match (Fig. 22). One team is attacking strongly.

Figure 23

Mark 22 'hits' on this rifle-range target (Fig. 23) by a moderately good marksman.

Figure 24

Mark 22 'hits' on this rifle-range target (Fig. 24) by a poor marksman.

(ii) Classify each of your dot patterns according to whether it is *clustered, regular* or *irregular* (see Exercise 2). Then describe the *factor* or *factors* which have influenced each pattern.

EXERCISE 4: Measuring distributions

Of course it is not very satisfactory just to guess whether a dot distribution is clustered, regular or irregular. It is much better to measure it.

Let us measure the amount of clustering in the patterns shown in Figures 25 and 26.

(i) For each of these diagrams:

(a) using the scale, measure (to the nearest half-kilometre) the distance from each point to its nearest neighbour;

(b) calculate the average nearest neighbour distance;

(c) state the type of distribution.

To decide on the type of distribution in each case, use the following key:

Average nearest neighbour distance	*Type of distribution*
3 km or over	Regular
Between 1 km and 3 km	Irregular
1 km or under	Clustered

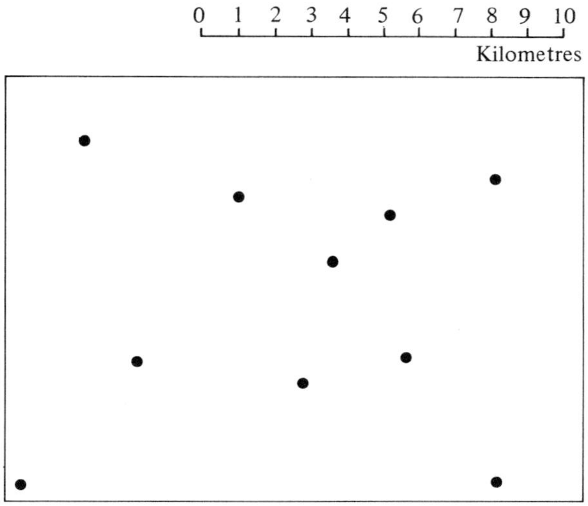

Figure 25

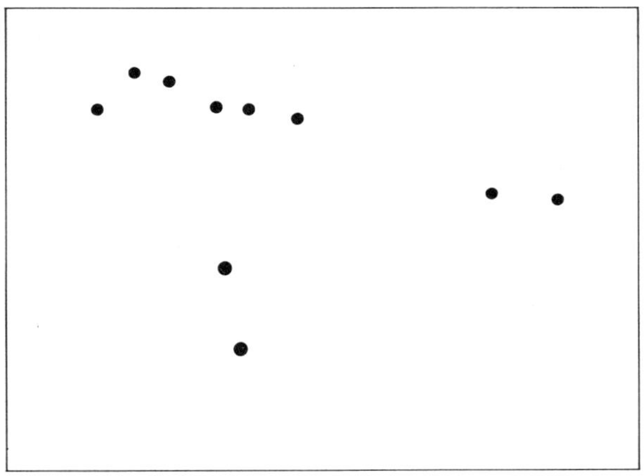

Figure 26

(ii) Now draw some more dot distributions for yourself. This key can be used for any other distribution of 10 dots within an area of the same size as the diagrams on this page.

EXERCISE 5: Types of clusters

In all kinds of point distribution patterns it is possible to recognize nearest neighbour 'clusters', whether or not the complete dot pattern is of the clustered type. It is often easier to understand a dot pattern by finding these 'clusters'.

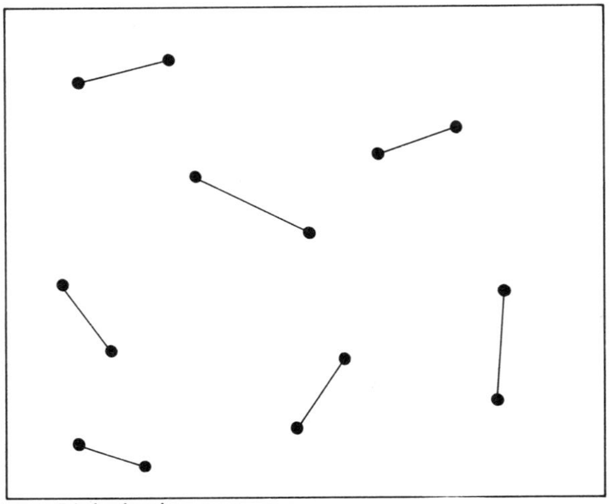

Linked pairs

Figure 27

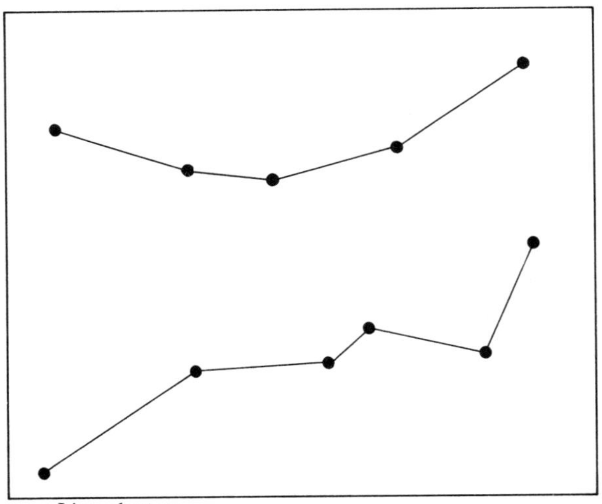

Line clusters

Figure 28

Three kinds of clusters will often be seen:

(i) *Linked pairs* (Figure 27)—each dot is its partner's nearest neighbour;

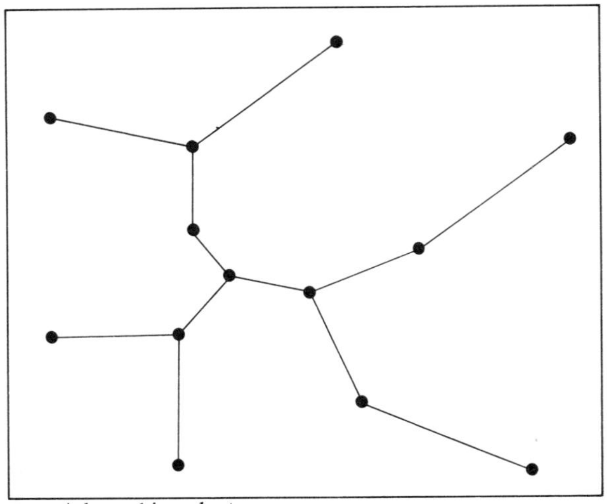

A branching cluster

Figure 29

(ii) *Line clusters* (Figure 28)—long single lines of points linked together by nearest neighbour lines;

(iii) *Branching clusters* (Figure 29)—nearest neighbour lines radiating outwards from a point near the centre of the cluster.

Try drawing some point patterns for yourself. Insert nearest neighbour lines. Then draw a ring round each cluster and label its type.

EXERCISE 6: A rural settlement pattern in the Yorkshire Dales

The map (Figure 30) shows the settlement pattern formed by 19 villages in a part of the Yorkshire Dales. This is an upland area with many steep slopes, and about 40 per cent of the area reaches an altitude of over 457 metres (1500 ft). The contour line for 457 metres is shown. The rivers Swale and Ure flow eastwards across the area.

(i) Make a tracing to show the locations of the 19 villages.

(ii) Draw nearest neighbour lines for each of the 19 villages.

(iii) Draw a ring round each nearest neighbour cluster and label its type.

(iv) Which types of clusters are most common here?
Suggest a reason.

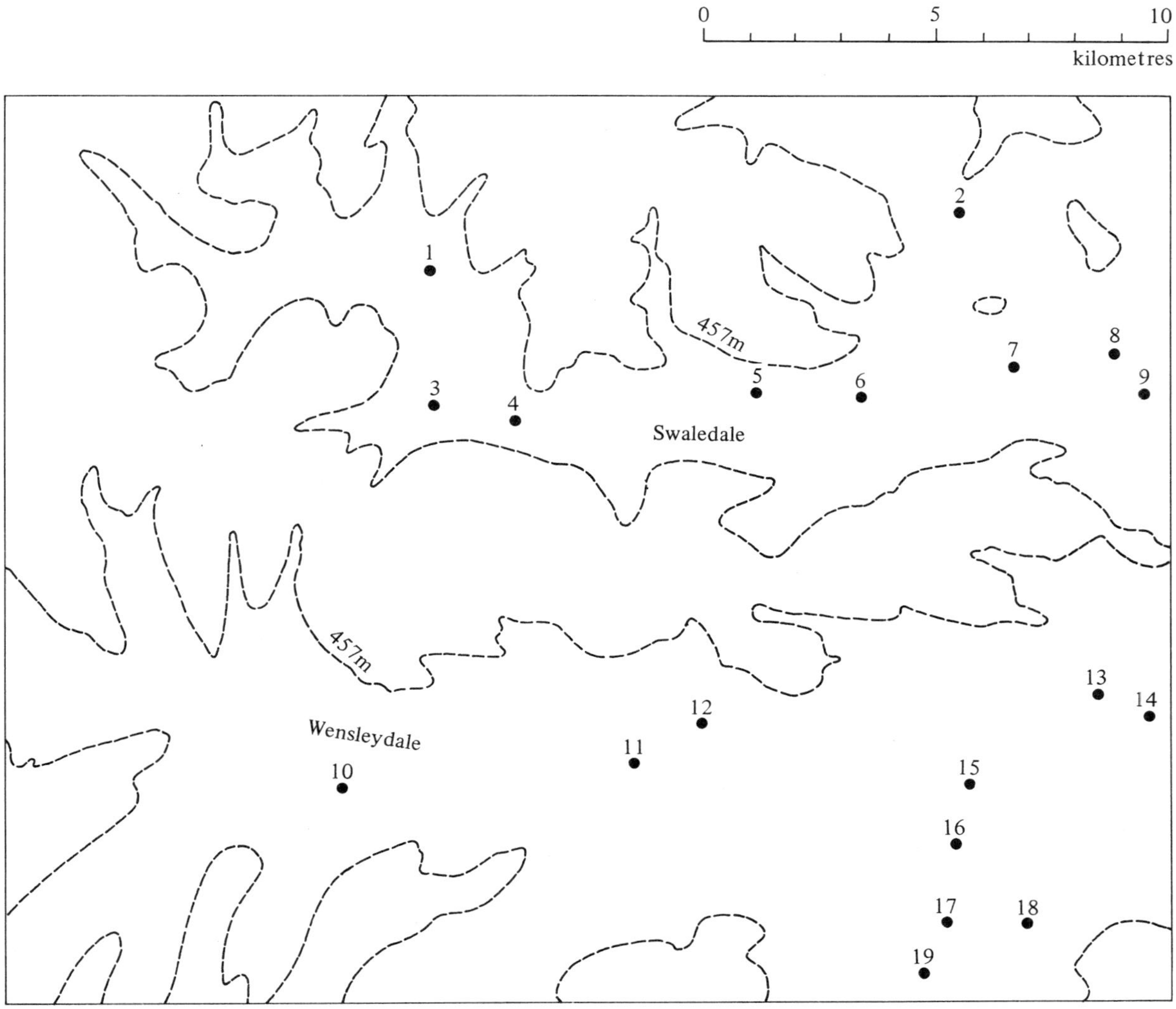

Figure 30

(v) On a rough piece of paper calculate the average nearest neighbour distance for these villages. Work to the nearest whole kilometre.
If your average nearest neighbour distance is 2 km or under the pattern may be called clustered. If it is 3 km or over it is regular.

(vi) What kind of pattern is this?
Suggest reasons why this settlement pattern is of this type.

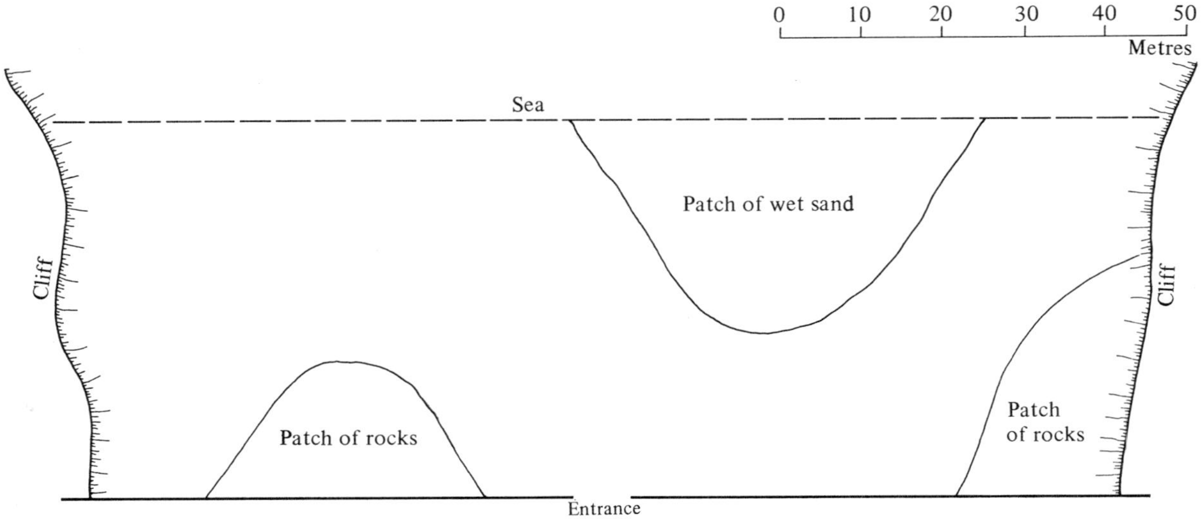

Figure 31

EXERCISE 7: Simulating a settlement pattern (1)

In the previous exercise we saw how a settlement pattern has been influenced by the relief of the land particularly.

Through the centuries settlement patterns have been created and altered as a result of people at various times making decisions. These decisions have been influenced partly by *physical factors* (altitude, steepness of slopes). Also, through the ages, people's attitudes have changed. The Anglo-Saxons choosing sites for farming settlements would have different attitudes from people choosing places for modern housing estates. These attitudes and values may be called *human factors.*

Let us try to understand the influence of physical and human factors by studying a type of settlement pattern which is created and destroyed in a single day in hundreds of different places round the British coast in summer. This is the pattern which is formed by family groups, with their deck chairs, on a beach on a fine, sunny day.

The outline of the beach is shown in Figure 31. No settlements have yet been created. As families begin to arrive and pass through the single entrance, the following factors influence their choice of settlement sites.

(a) *Physical factors.* They wish to settle on a patch of clean, dry sand. Areas of wet sand, rocks etc. are not desirable.

(b) *Human factors.* They wish to settle at the shortest possible distance from the beach entrance because father tends to grumble if he has to carry the deck chairs too far. But mother will not accept a location which is less than 10 metres away from a neighbouring group because she likes peace and quiet.

Make a tracing of the map of the beach, not forgetting to include the scale. Now mark dots where you think the first 12 family groups will settle. Use the scale to make sure that they are far enough apart. Work in pencil.

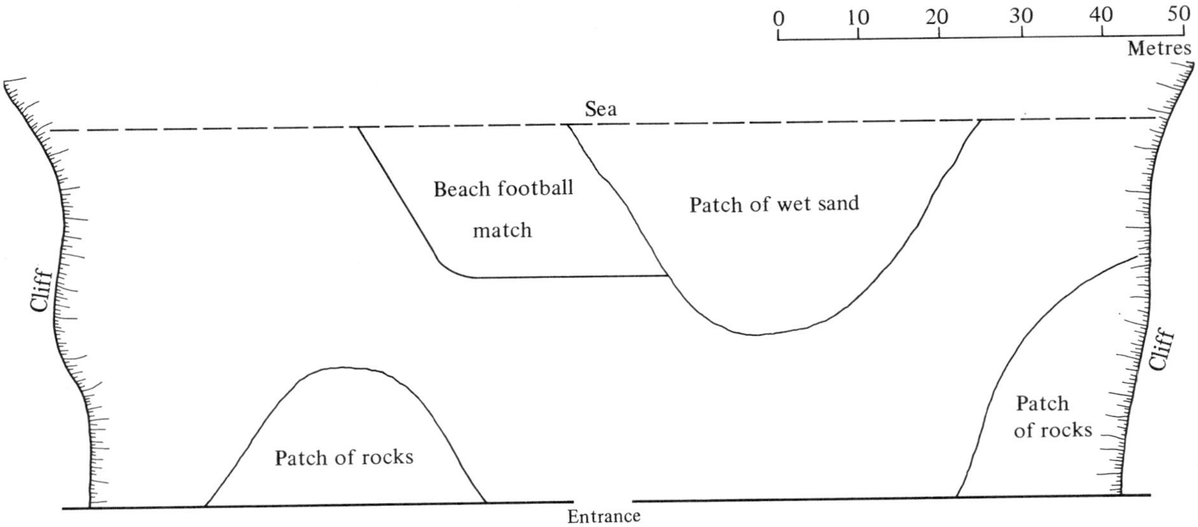

Figure 32

EXERCISE 8: Simulating a settlement pattern (2)

As time goes by an additional *human factor* develops. This is a beach football match in the area shown in Figure 32. Some nearby family groups may find this rather annoying and decide to move.

(i) Mark the location of the football match on your beach map. Then:
(a) rearrange your original 12 settlements to take account of this new factor;
(b) add 12 more family groups to the settlement pattern, considering all the factors which have so far operated.

Now 16 more families arrive in turn. Some of these families may be faced with a considerable walk to find a quiet place. Some of them therefore may decide that, rather than walk a long distance, they will settle between existing groups, but as far away from them as possible.

(ii) Mark 16 more dots on your beach map to show where these 16 families are likely to settle.

Your map now shows the complete settlement pattern of 40 families. Compare and discuss your pattern with the ones produced by others in your class.

Unit IV: Market areas (2)

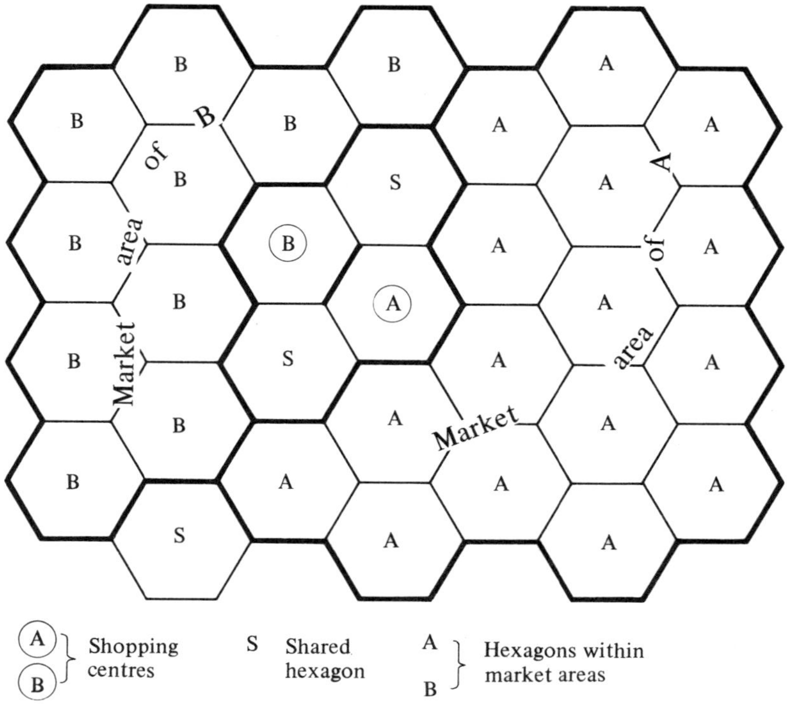

Figure 33

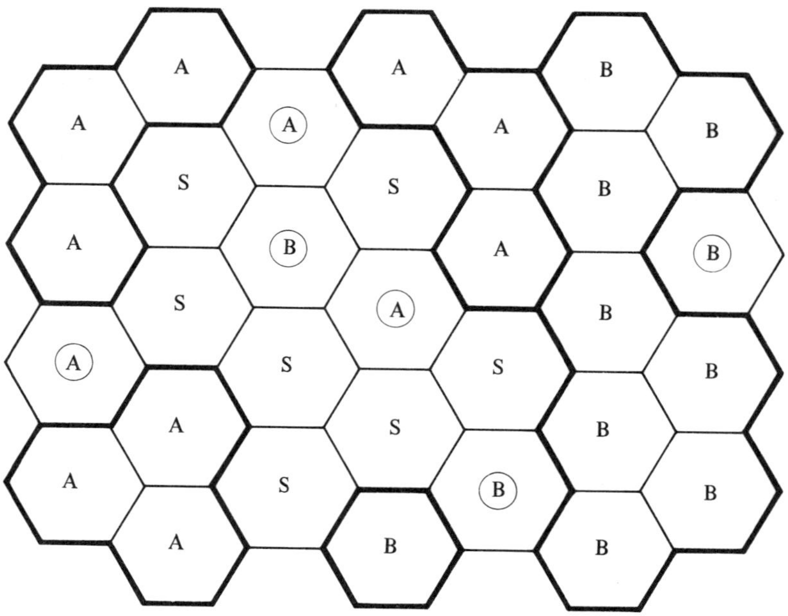

Figure 34

EXERCISE 1: Shopping centre competition game (1)

This is a game similar to the one about competition between shops in a village (page 9).

The patterns of hexagons in Figures 33 and 34 represent part of a town. Each hexagon contains the homes of the same number of people, who all do their shopping at the shopping centre which is nearest to the hexagon in which they live.

The two players build shopping centres alternately until each has built three. Each time a shopping centre is built it takes over the whole of one hexagon. The object of the game is for each player to locate his shopping centres so that their market areas contain as many hexagons as possible. Figure 33 shows the state of the game when each player has built one shopping centre. Count how many hexagons lie within each market area. *Do not count:* (i) hexagons containing shopping centres; (ii) 'shared' hexagons which are an equal distance from rival shopping centres (these are labelled S).

Figure 34 shows the completed game. Count up the hexagons and find the final score.

EXERCISE 2: Shopping centre competition game (2)

Now play the game explained in the previous exercise using blank diagrams.

Toss a coin to decide which player plays first.

Each player alternately locates a shopping centre, until each has three.

Scoring
At the end of each game: (i) mark all 'shared hexagons' S; (ii) label all hexagons which are within each player's market areas.

In scoring *do not count*: (i) hexagons occupied by shopping centres; (ii) 'shared' hexagons.

EXERCISE 3: Shopping centre competition game (3)

It is very rare, of course, for the built-up area of a town to form a regular shape like that used in the game so far. Try the game now using an irregular shape. The rules of play and scoring are exactly the same as before.

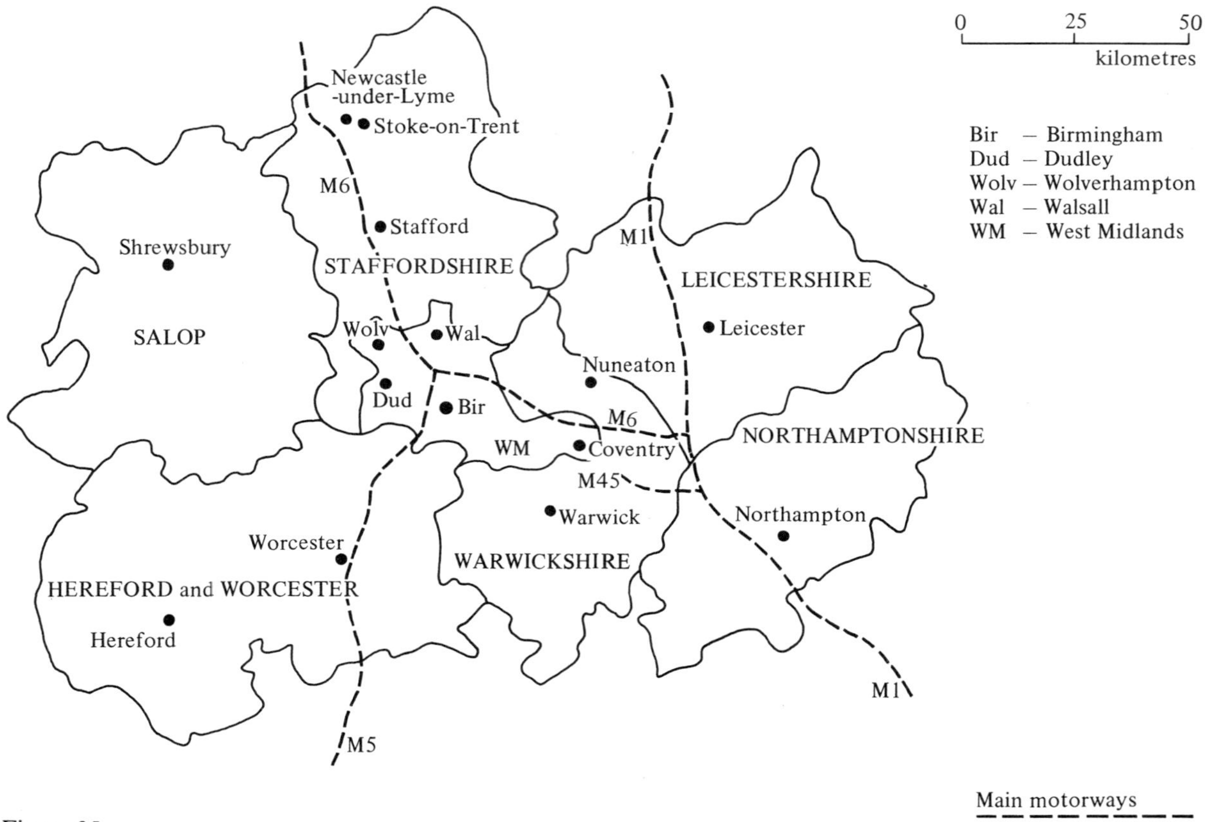

Figure 35

EXERCISE 4: Locating shopping centres in the English Midlands

You are given the task of planning the location of a chain of modern shopping centres in the Midland counties shown on the sketch map (Figure 35). There are to be 20 shopping centres and, in order to be profitable, each needs a market area with a population of at least 300 000.

Study the sketch map and the table of population totals.

Population in 1974 (thousands)

County	*Main Towns*
Hereford & Worcester (586)	Worcester (75)
	Hereford (47)
Leicestershire (830)	Leicester (287)
Northamptonshire (496)	Northampton (138)
Salop (354)	Shrewsbury (84)
Staffordshire (991)	Stoke-on-Trent (258)
	Newcastle under Lyme (122)
	Stafford (114)
Warwickshire (470)	Warwick (111)
	Nuneaton (112)
West Midlands (2780)	Birmingham (1086)
	Dudley (299)
	Walsall (271)
	Wolverhampton (268)
	Coventry (335)

(i) Make a tracing of the county boundaries and the motorway pattern shown on the sketch map. Then mark 20 dots on your map to represent your choice of shopping centre locations. Try to draw coloured lines to show the market areas of different shopping centres.

(ii) Compare and discuss your pattern with those of your friends.

Section B
Lines in Geography

Unit I: Lines as routes

EXERCISE 1: Mazes

(i) Study the maze in Figure 36.

(a) From which of the entrances (A, B, C or D) is there a direct route to X without crossing a line?

(b) It is possible to enter the maze at one of entrances A, B, C and D, travel through the maze without crossing a line, and leave the maze at a *different* entrance. Which two entrances are used?

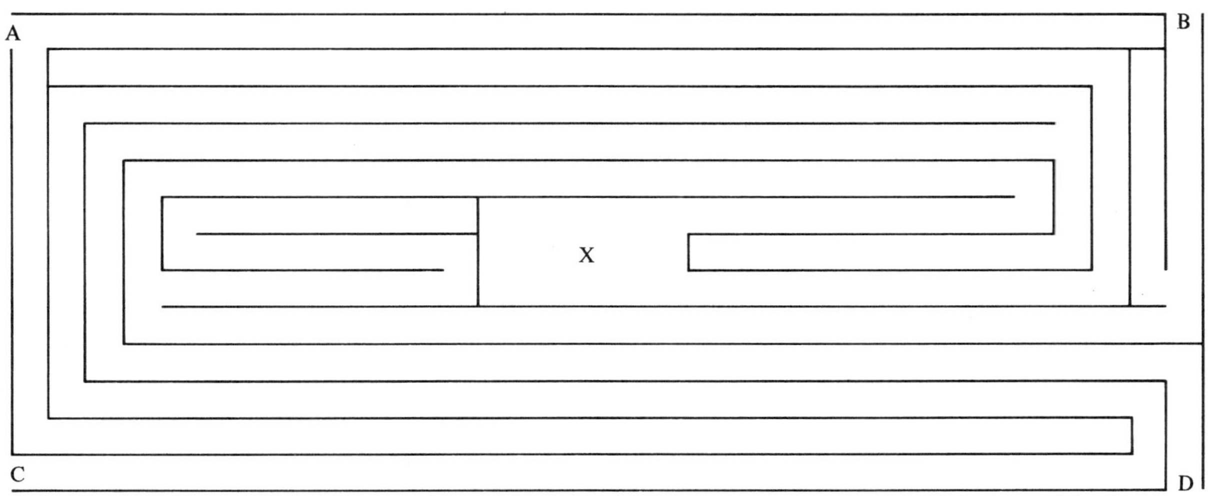

Figure 36

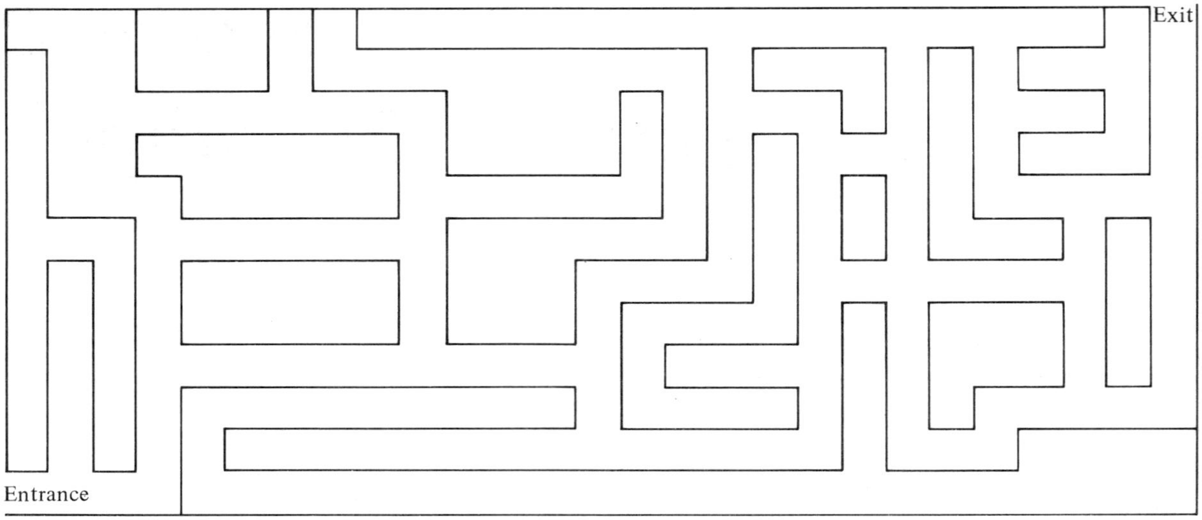

Figure 37

(ii) Study the maze in Figure 37.

(a) In this maze there are several different ways of travelling from the entrance to the exit. Find the shortest of these routes.

(b) Write short instructions such as you might give to a person who wanted to take this shortest route through the maze but did not possess a map.

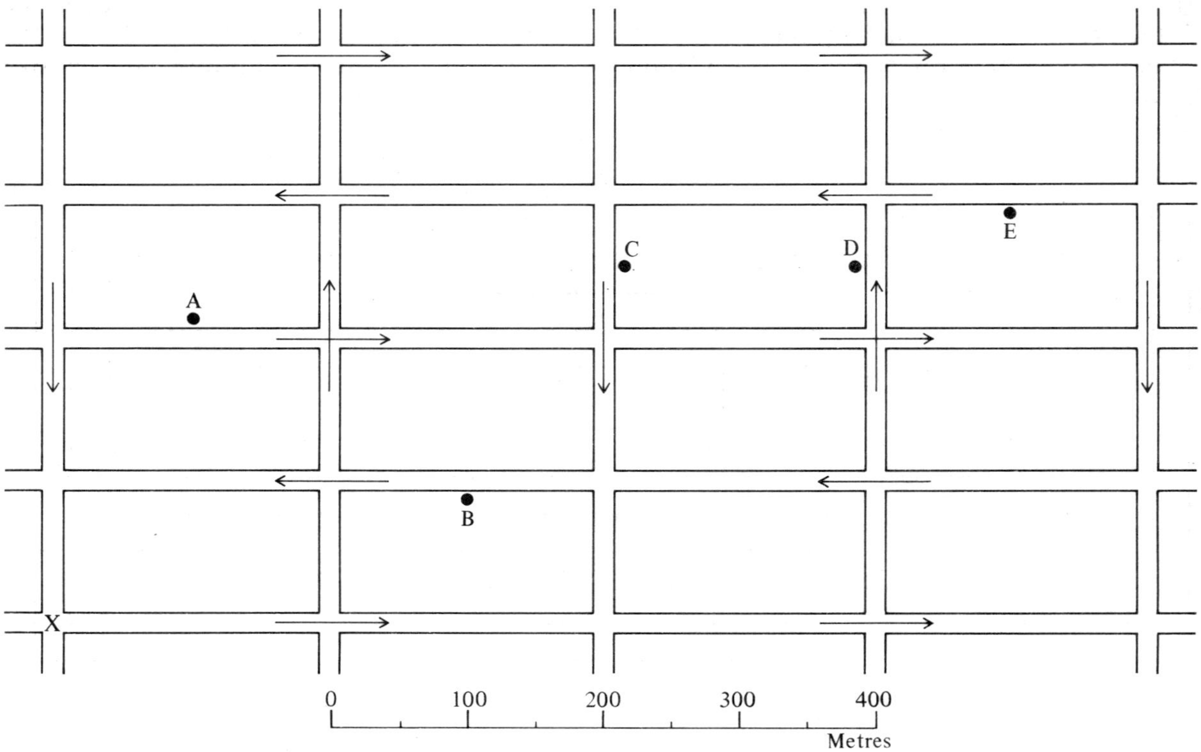

Figure 38

EXERCISE 2: One-way streets in a gridiron street plan

Figure 38 shows a gridiron pattern of one-way streets such as might be found in an American city.

(i) Beginning at the point marked X, find the shortest route by car to each of A, B, C, D and E separately. What is the approximate distance travelled in each case?

(ii) In each of the following cases, state which of the two journeys is longer: (a) A to B or B to A; (b) B to C or C to B; (c) A to E or E to A; (d) B to D or D to B.

(iii) A motorist at the point marked X wishes to visit B, C and D and then return to X. In which order should he visit these places so as to travel the shortest possible distance?

EXERCISE 3: One-way streets in a hexagonal street plan

(i) Figure 39 shows a one-way street pattern laid out in the form of hexagons. What main advantage for traffic flow has this system compared with the rectangular pattern shown in the previous exercise?

(ii) Beginning at the point marked X, find the shortest possible journey by car to each of places A, B, C, D and E separately. What is the approximate distance travelled in each case?

(iii) Give the approximate length of the shortest possible journey by car in each of the following cases: (a) A to B; (b) D to E; (c) C to E; (d) E to B; (e) B to D.

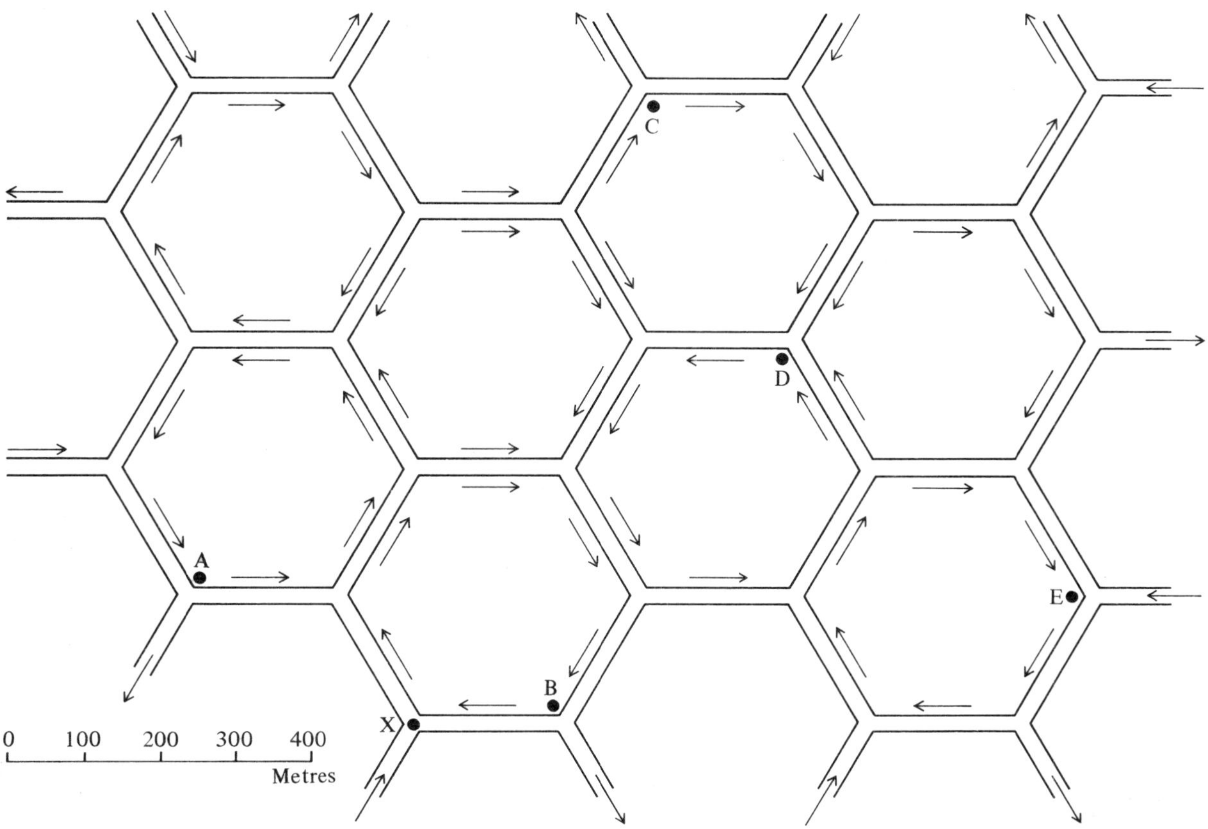

Figure 39

EXERCISE 4: Planning a bus route (1)

Figure 40 shows some of the streets in the central shopping district of a town. It is proposed to introduce a 'circular tour' bus service for the convenience of shoppers. The bus has to start at X and also it has to end its journey at X. In its 'round trip' it has to pass along each of the streets *once* but *not more than once*. But it may pass through any street junction any number of times.

(i) List the street numbers in the order in which the bus passes along them.

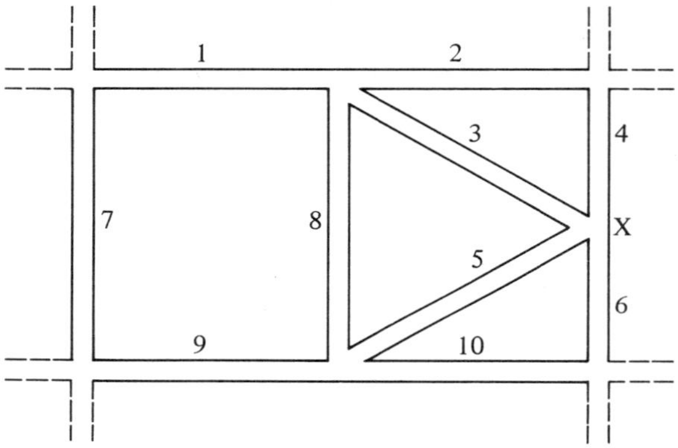

Figure 40

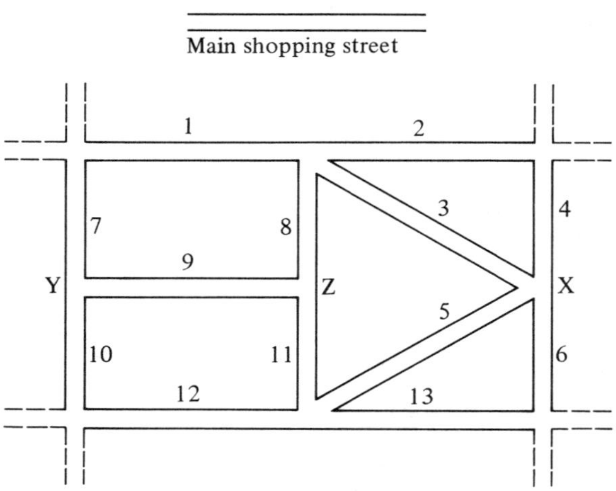

Figure 41

A new road (YZ) is now built, as shown in Figure 41.

(ii) Try to re-organize the bus route, starting and ending at X. Remember that the bus must pass along all the streets once and once only. What is your conclusion?

(iii) Let the bus begin its journey at Y. Where does its journey now end?

(iv) How are the street junctions at Y and Z different from those in the remainder of the map?

If you have answered this question correctly it will help you in the planning of bus routes in the next exercise.

EXERCISE 5: Planning a bus route (2)

(i) Imagine that you live in the town whose road pattern is shown in Figure 42. A party of foreign visitors wishes you to explain to them the details of the geography of the town. Because time is short you decide first of all to take them by bus along all the roads shown on the map.

Beginning at any road junction, plan a route which passes along each road *once and once only*. Make a tracing of Figure 42 and number the roads in the order in which the bus should pass along them. The route must end at the junction at which it started.

(ii) The mayor of a neighbouring town whose road pattern is shown in Figure 43 hears of the success of your trip, so he invites you to plan a similar route for his own town.

When you study the road pattern in Figure 43 you decide that you will have to be very careful which junction you choose as a starting point. You also realize that you cannot end the trip at the junction at which you started. Make a tracing of Figure 43 and number the roads in the same way as you indicated the route in your own town.

Note: It may help to refer back first of all to Exercise 4.

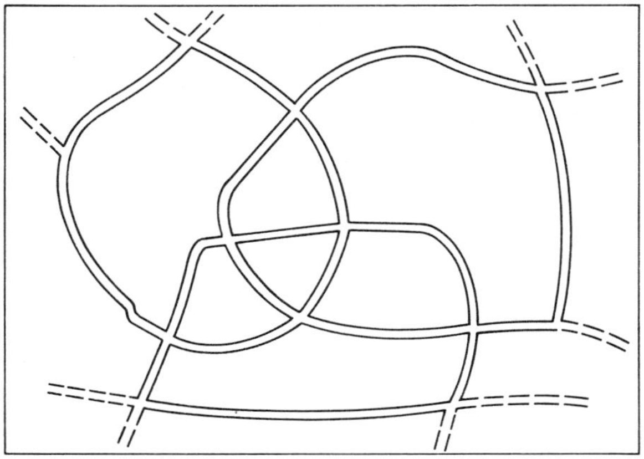

Figure 42

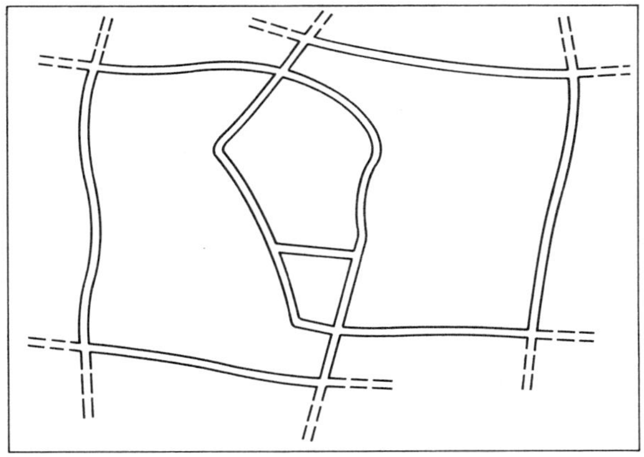

Major Road

Figure 43

Unit II: Lowest-cost routes

EXERCISE 1: The effect of barriers

In the areas shown in Figures 44 to 46 we are trying to plan the lowest-cost route for a railway or a road from the Port to the City. Any route must pass through the *centres* of the hexagons. The number marked in each hexagon shows the cost of building the road or railway through that hexagon.

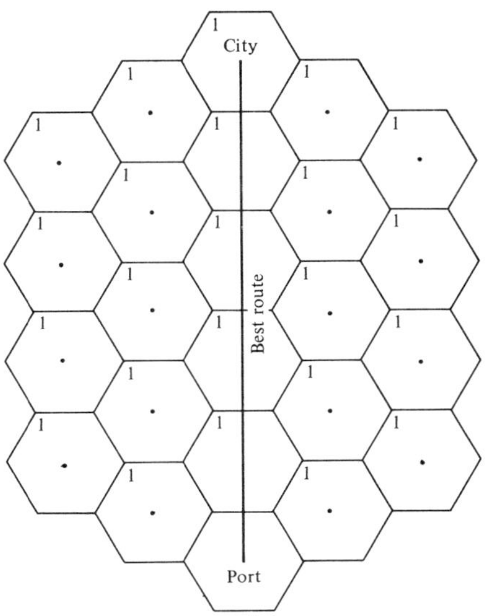

Figure 44

In Figure 44 the shortest route is clearly the cheapest. The total cost is 5 units.

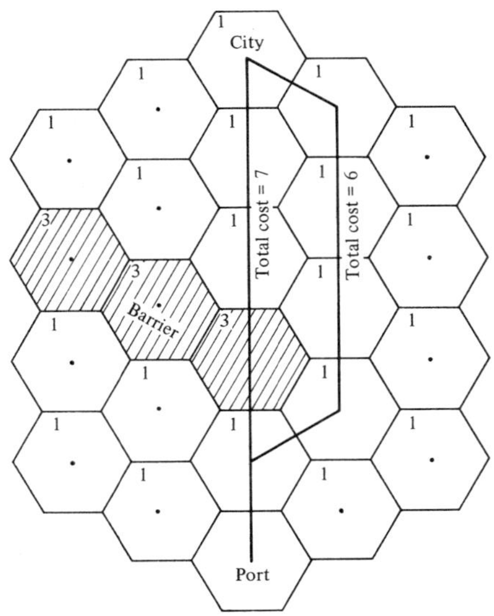

Figure 45

In Figure 45 we have now introduced a barrier, which might be a mountain range or a river, where the cost of building is much higher. You will see that in this case it is cheaper to bypass the barrier rather than to go straight across it.

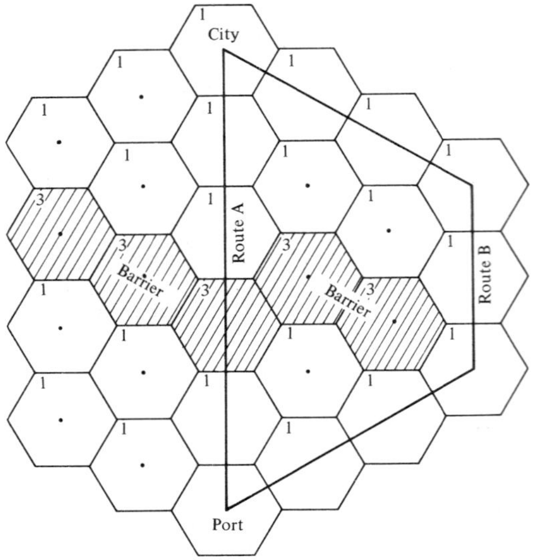

Figure 46

The barrier in Figure 46 is much longer, and the lowest-cost route is now straight through the barrier. The long diversion costs more than the tunnelling or bridge-building needed to pass straight through the barrier.

(i) Calculate the cost of each route in Figure 46.

(ii) Find examples in an atlas of road and railway routes: (a) going straight across barriers; (b) taking a longer route so as to avoid a barrier.

EXERCISE 2: The effect of economic development

Another consideration in the planning of a transport route might be the chance of attracting traffic so as to earn revenue to pay for the building and upkeep costs.

A railway, for example, might be routed through a developing farming or mining area from which an export trade could arise, with products being carried by rail. Or there may be a town which would generate passenger and goods traffic. It may therefore reduce costs in the long run for the railway to pass through such areas.

In Figures 47 and 48, such traffic producing areas are given minus numbers (−1, −2, etc.) which means that they produce a profit for the transport route. If a route passes through one of these areas this number is subtracted from the costs. The costs for **Route** A in Figure 47 are: 1 + 1 + 1 + 1 − 1 + 1 + 1 + 1, making a total of 6.

(i) State the total cost for: (a) **Route** B; (b) Route C; (c) Route D.

(ii) What is the total cost for the lowest-cost route from the Port to the City in Figure 48, passing through as many of the profit-making areas as you wish?

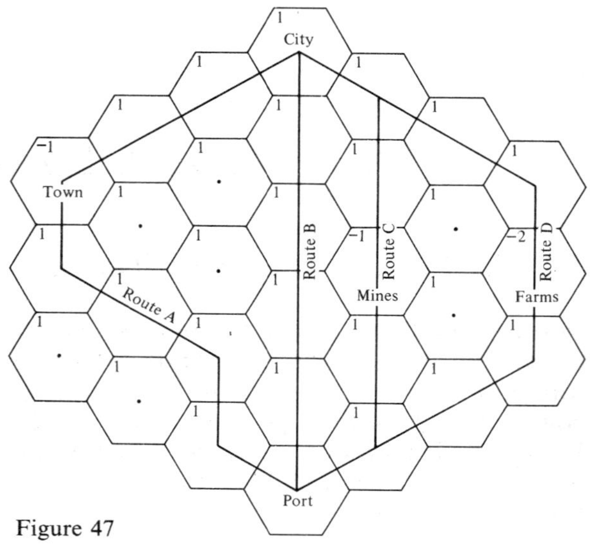

Figure 47

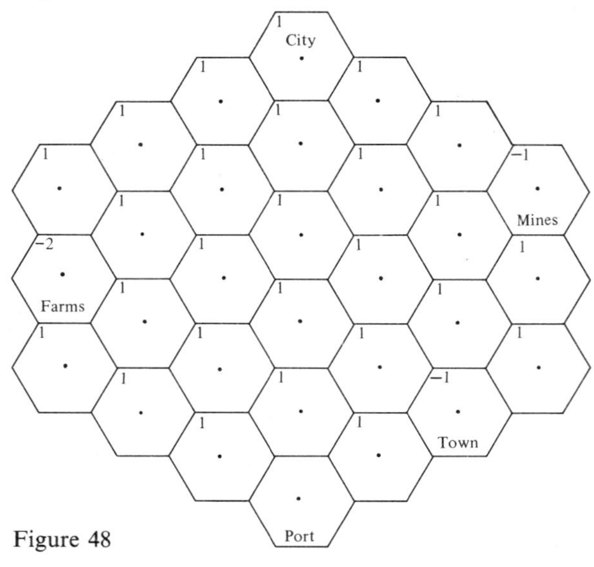

Figure 48

EXERCISE 3: Lowest-cost route-finding game

Here is a game which you can play by yourself or as a competition with some of your friends.

As in the earlier exercises, the object is to plan the lowest-cost route from the Port to the City. As you can see, there are several barriers which increase costs, and also there are several areas of farms, mines and towns which reduce costs.
To play this game use the map on the separate sheet.

EXERCISE 4: The influence of relief on routes in the Scottish Highlands

Figure 49 shows an area in the southern part of the Highlands of Scotland. The dotted line is the contour for 305 metres (1000 ft). The double lines round the margin of the map are the main roads which enter this area from other parts of Scotland. The road pattern within this area is strongly influenced by relief.
Lay a sheet of tracing paper over this map and draw lines to show where you think the main roads of this area run. All the settlements named on the map are situated on main roads.
You can check your road pattern from a road atlas.

Figure 49

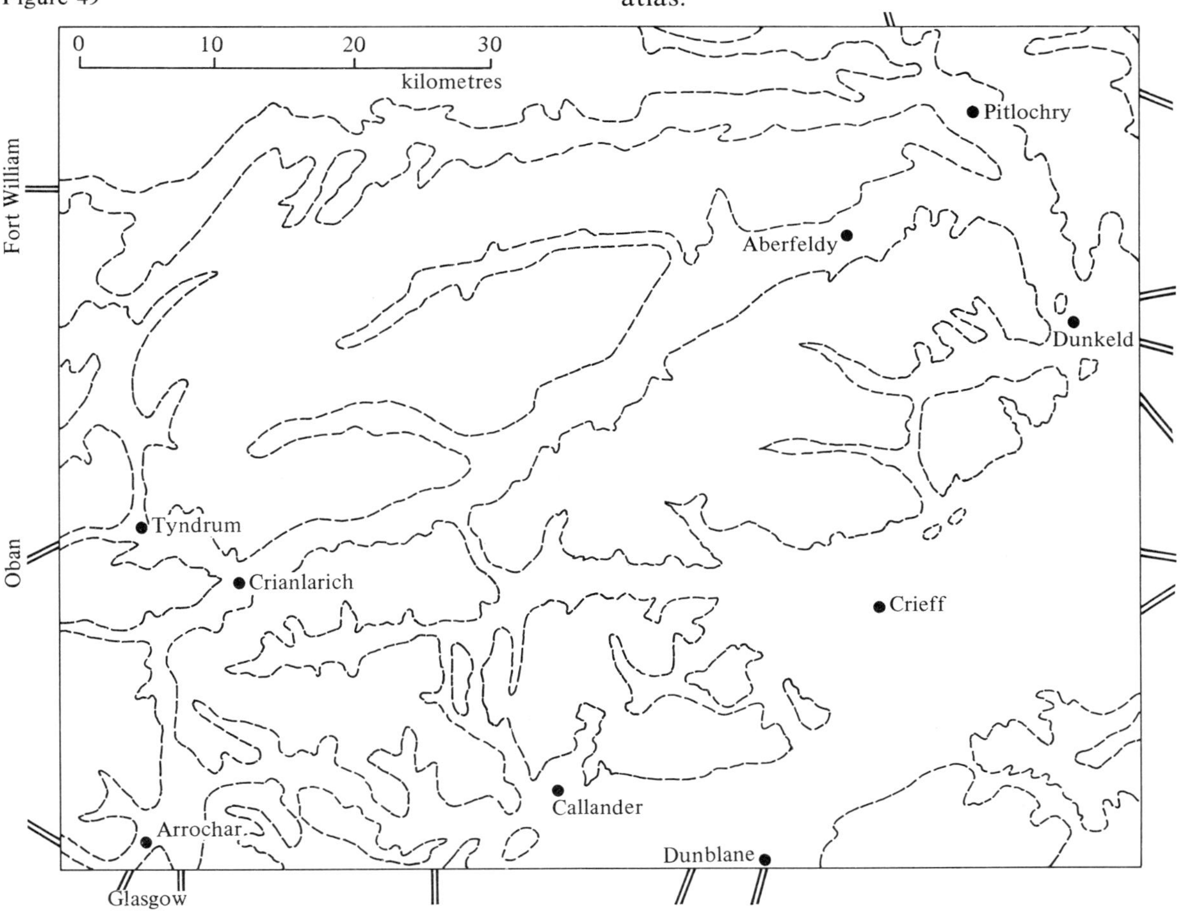

Unit III: Sea transport routes

EXERCISE 1: Barriers to sea routes

Just as areas of mountains are often barriers to roads and railways, so long peninsulas act as barriers to sea routes.

Figure 50 shows a sea area with a peninsula extending down its centre.

(i) In the first place, imagine that the ship canal has not yet been made.

Make a copy of the matrix labelled A and in it write the shortest distance from each of ports A, B and C to each of ports X, Y and Z. Your routes must pass through the dots at the centres of the hexagons. In calculating distance, count every hexagon the route enters except the port from which it starts. The route from C to Z has been done for you.

(ii) Now copy Matrix B and write in the shortest distance from port to port using the ship canal for all routes.

(iii) Make a copy of Matrix C and in it write the distance saved by using the ship canal for each route.

(iv) Which sea route is longer via the ship canal than by sea all the way?

(v) Which ports gain nothing from the use of the ship canal?

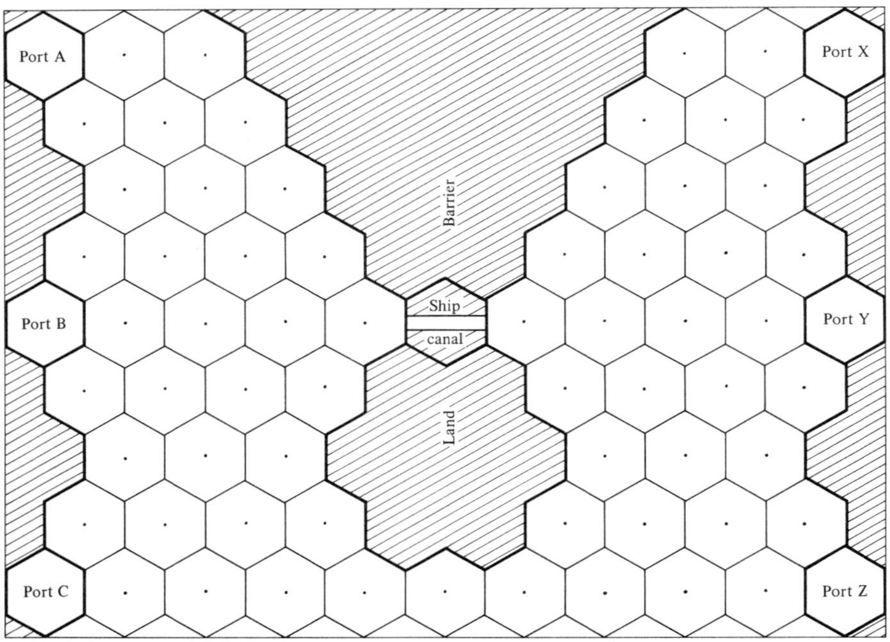

A
Distance without using the canal

	X	Y	Z
A			
B			
C			10

B
Distance using the canal in all cases

	X	Y	Z
A			
B		10	
C			

C
Distance saved by using the canal

	X	Y	Z
A			
B			
C			

Figure 50

EXERCISE 2: The world's great peninsulas

Sea routes on a world scale are much influenced by long, north–south peninsulas. You may be more familiar with the map in Figure 51 than that in Figure 52, yet in many ways Figure 52 is better. It shows how the northern continents form a ring around the north pole from which long peninsulas like South America and Africa extend southwards. Also, in Figure 52, all distances measured in a straight line from London are true to scale. The map in Figure 51, on the other hand, is very much distorted.

(i) In Figure 51 which of the named cities seems nearest to London?

(ii) Which of the named cities actually is nearest to London? (See Figure 52.)

(iii) Which two continents have been omitted from both maps?

(iv) The reason for this is that they are both difficult to show in Figure 52. Why is this?

(v) Describe the shortest sea routes from London to San Francisco, Cape Town, Colombo and Tientsin as they were before any canals were made. No sea route can cross the Arctic Ocean.

(vi) Describe the shortest sea routes from London to the other four cities using the Suez and Panama Canals if necessary.

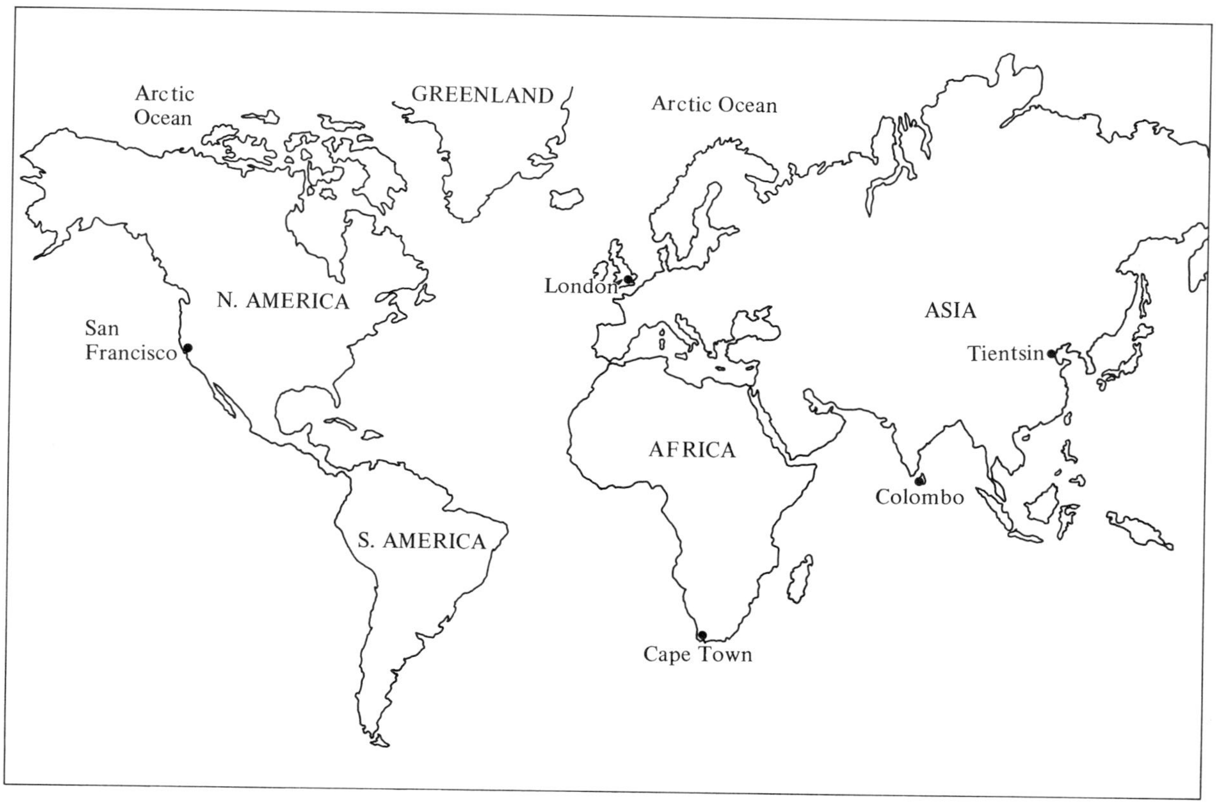

Figure 51

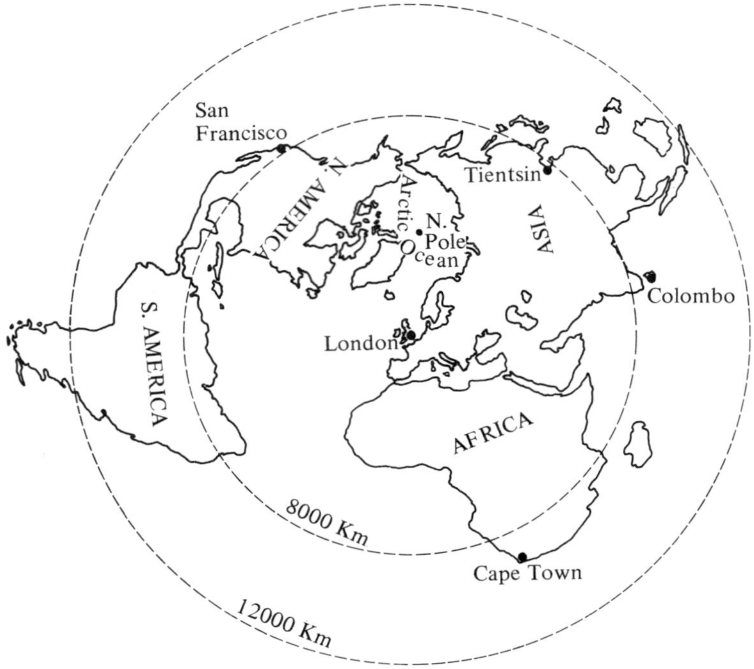

Figure 52

EXERCISE 3: The Suez and Panama Canals

(i) *The Panama Canal*
The Panama Canal is a way of bypassing the long north–south barrier of South America instead of passing round Cape Horn.

On the map in Figure 53 four ports on the west coast of the Americas and four ports on the east coast are marked and named. Copy the matrix and, in it, write P wherever the shortest possible sea journey between a pair of ports is through the Panama Canal. Draw a sketch map showing these sea routes which pass through the Panama Canal.

	West coast ports: Los Angeles	Guayaquil	Callao	Valparaiso
East coast ports: New Orleans				
Recife				
Rio de Janeiro				
Buenos Aires				

Figure 53

(ii) *The Suez Canal*
The Suez Canal is a way of bypassing the long north–south barrier of Africa, instead of passing round the Cape of Good Hope, near Cape Town.

Copy the matrix in Figure 54 and, in it, write S wherever the shortest possible sea journey between a pair of ports is through the Suez Canal. Draw a sketch map showing these sea routes which pass through the Suez Canal.

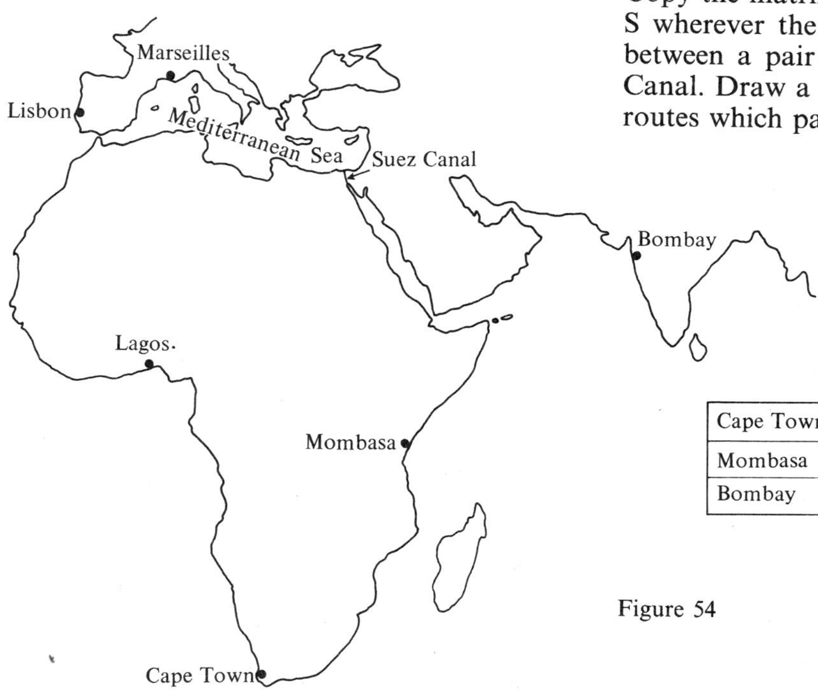

	Lagos	Lisbon	Marseilles
Cape Town			
Mombasa			
Bombay			

Figure 54

Unit IV: Transport costs along routes

EXERCISE 1: Locating an industry (1)

We are now going to look at the ways in which routes may influence where industries are.

Imagine that there is a certain industry which uses two raw materials (which we shall call A and B) to produce a product which is then sold at a town which we shall call the 'market'. A businessman is wondering where to set up his factory. He realizes that he will have to bring the two raw materials to his factory, wherever he locates it, and that he will have to send his product to the market, where it can be sold.

Figure 55 shows where the two raw materials are found, and where the market is situated. He decides to set up his factory wherever the costs of transport are lowest.

The industry uses 1 tonne of raw material A and combines this with 1 tonne of raw material B. This gives 2 tonnes of product to be sent to the market. It costs 1p to transport each

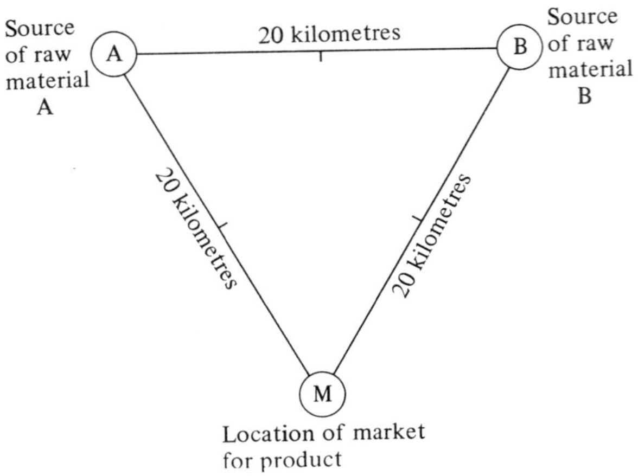

Figure 55

tonne of raw material or product a distance of 1 kilometre (i.e. 20p for 20 kilometres).

Let us work out what the businessman's transport costs will be at three different locations.

	Cost of transport			
Location	*Raw material* A *to factory*	*Raw material* B *to factory*	*Product to market*	*Total*
Source of raw material A	Nil	1 tonne for 20 km = 20p	2 tonnes for 20 km = 40p	60p
Source of raw material B	1 tonne for 20 km = ——	——	2 tonnes for 20 km = ——	60p
Market (M)	1 tonne for 20 km = ——	1 tonne for 20 km = ——	Nil	——

(i) Several blank spaces have been left in this table. Make a copy of the table and fill in the blank spaces.
(ii) Where should the businessman set up his factory?

EXERCISE 2: Locating an industry (2)

Let us now study the effects of different patterns of routes.

In both of the exercises on this page, remember that:
(a) 1 tonne of raw material A added to 1 tonne of raw material B produces *2* tonnes of product;
(b) transport costs for both raw materials and product are 1p per tonne per kilometre.
(i) First calculate the transport costs in Figure 56 and put them in a table like the one below.

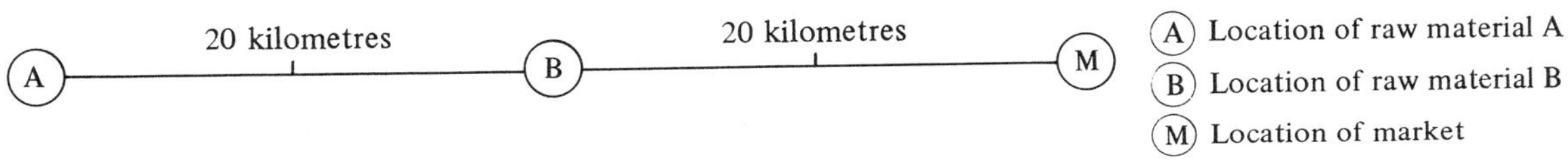

Figure 56

Location	*Raw material* A *to factory*	*Cost of transport* *Raw material* B *to factory*	*Product to market*	*Total*
A	Nil	20p	——	——
B	——	——	——	——
M	——	——	——	——

Lowest-cost location: ——————

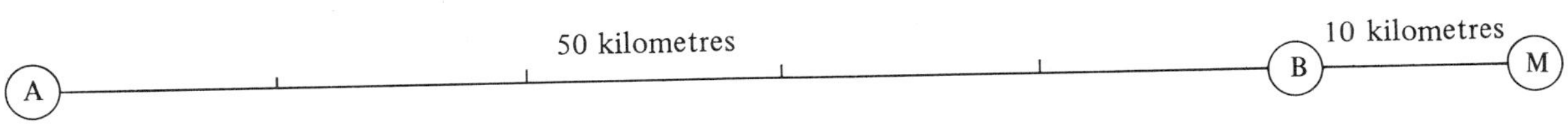

Figure 57

(ii) In Figure 57, the spacing of the raw materials has been changed to see what difference this makes. Calculate the transport costs and make a table like the one you have made for Figure 56.

(iii) You will see that the market has a strong attraction for this sort of industry in which all or nearly all of the raw material goes into the finished product and little or no raw material is wasted. Name some industries which you think may be of this type.

EXERCISE 3: The wastage factor

In many industries much waste is produced during the manufacturing process. In saw-milling, for example, much of the tree is of little value and cannot be used. In such industries, 2 tonnes of raw materials may only produce 1 tonne of product instead of 2 tonnes. 1 tonne of the raw materials may have to be thrown away as waste. Let us now study the effect of this 'wastage factor' on transport costs and the location of industry.

In both of these exercises (a) 1 tonne of raw material A added to 1 tonne of raw material B produces *1* tonne of product; (b) transport costs of both raw materials and product are 1p per tonne per kilometre.

(i) Calculate the transport costs in Figure 58 and put them in a table like the one below.

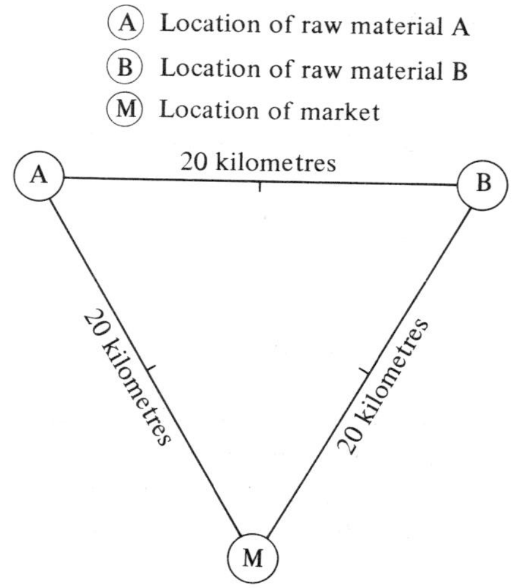

Figure 58

Location	*Raw material* A *to factory*	*Cost of transport* *Raw material* B *to factory*	*Product to market*	*Total*
A	____	____	____	____
B	____	____	____	____
M	____	____	____	____

Lowest-cost location: ____________

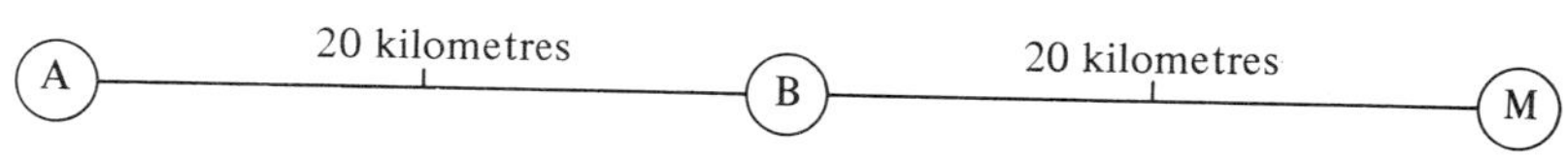

Figure 59

(ii) Now make another table to show the transport costs for Figure 59, to see if a different pattern of the location of raw materials and product makes any difference.

You will now see that the market location has lost some of its importance. After all, there is not much point in transporting raw materials to the market if so much of them is going to be wasted. It makes more sense to manufacture the product near the raw material, so that the waste material is never transported.

EXERCISE 4: An intermediate location

It is by no means certain that an industry will always be located at either the raw material or the market. Sometimes its lowest-cost location may be somewhere between the raw materials and the market. The oil-refining industry, for example, is often located well away from both raw materials and markets, as for example, at Milford Haven in South Wales.

Figure 60 shows the location of two raw materials (A and B) and the market (M). X is situated at a transport junction.

(i) In this exercise: (a) 1 tonne of raw material A added to 1 tonne of raw material B produces *1* tonne of product; (b) transport costs of both raw materials and product are 1p per tonne per kilometre.

Find the lowest-cost location using a table like the one below.

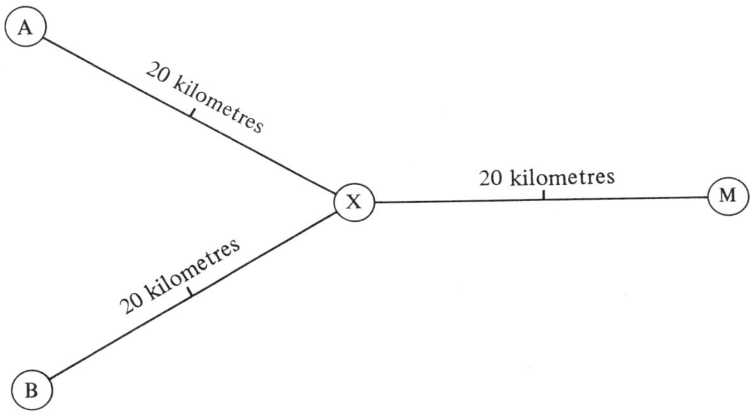

Figure 60

	Cost of transport			
Location	*Raw material* A *to factory*	*Raw material* B *to factory*	*Product to market*	*Total*
A	___	___	___	___
B	___	___	___	___
M	___	___	___	___
X	___	___	___	___

Lowest-cost location: ____________

(ii) Now suppose that an even greater amount of raw material is wasted during manufacture, so that the 2 tonnes of raw materials only produce $\frac{1}{2}$ tonne of product. Make a table like the one above and find the lowest-cost location.

You will see that the importance of an intermediate location depends partly upon the direction of the transport routes and partly upon the amount of raw material that is wasted during manufacture (wastage factor).

EXERCISE 5: Simulating the distribution of industry

Let us now try to build up a pattern of industrial location in the imaginary country shown in Figure 61. This country has three different industries which we shall call P, Q and R. These industries have the characteristics given below.

INDUSTRY P
Uses 1 tonne of raw material A and 1 tonne of raw material B to make *2* tonnes of product.

INDUSTRY Q
Uses 1 tonne of raw material C and 1 tonne of raw material D to produce *1* tonne of product.

INDUSTRY R
Uses 1 tonne of *each of* raw materials A, C and D to make *1* tonne of product.

(i) Calculate the lowest-cost locations for these industries. All transport costs are 1p per tonne per kilometre.
(ii) Copy Map B and mark on it P, Q and R at their lowest-cost locations. Choose between A, B, X and M for Industry P, between C, D, Y and M for Industry Q, and between A, C, D, X, Y and M for Industry R. If there are two equal lowest-cost locations, locate the industry at *both* of them.
(iii) Draw lines on your map to show the movements of materials and products. Use a different colour for each industry.

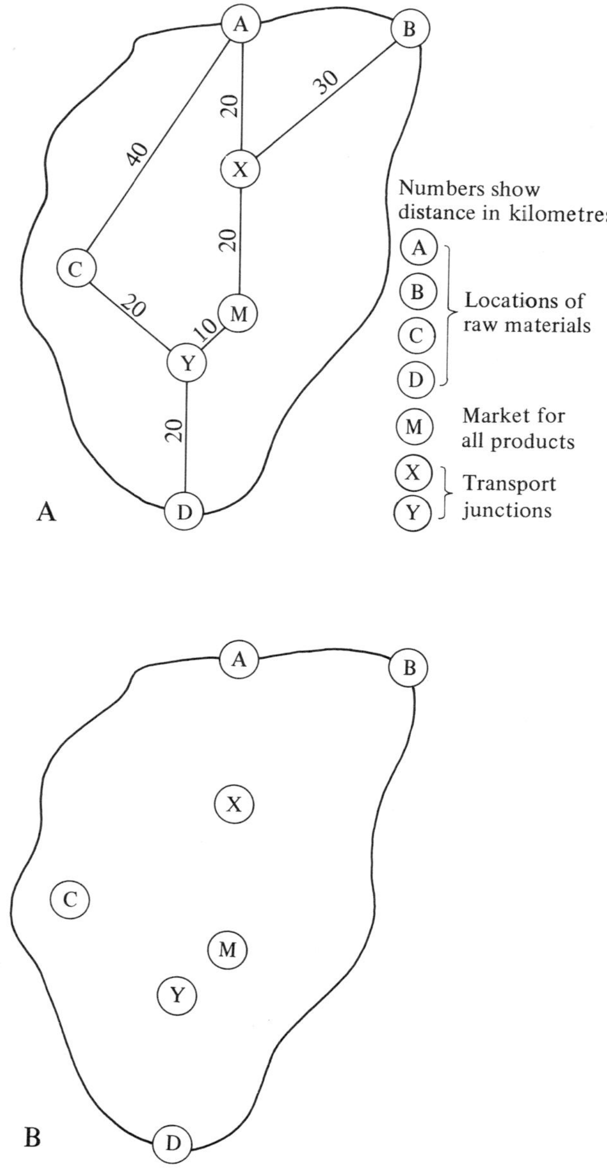

Figure 61

Section C
Networks in Geography

Unit I: Networks and movement

EXERCISE 1: River networks

The pattern which is formed by a branching set of lines is called a network. One of the commonest is a river drainage network as shown in Figure 62. This is the pattern formed by a river and its tributaries as they drain water from an area of land to the sea.

Each river network is separated from its neighbours by higher land called a *watershed* or a divide. It is sometimes called a divide because it 'divides' the water which flows into one network from that which flows into another. Near the watershed there are a number of small *tributary streams* (labelled T in Figure 62). These flow downhill away from the watershed and join together forming *confluences* (labelled C in Figure 62). Gradually more and more of the streams join together until a large stream (the trunk stream) is formed. This carries water from the whole drainage basin into the sea.

(i) Figure 63 shows a number of river networks. Place a piece of tracing paper over this map and draw lines to show the watersheds.

(ii) Figure 64 shows a pattern of watersheds. Place a piece of tracing paper over this map and draw river networks to fit the watersheds.

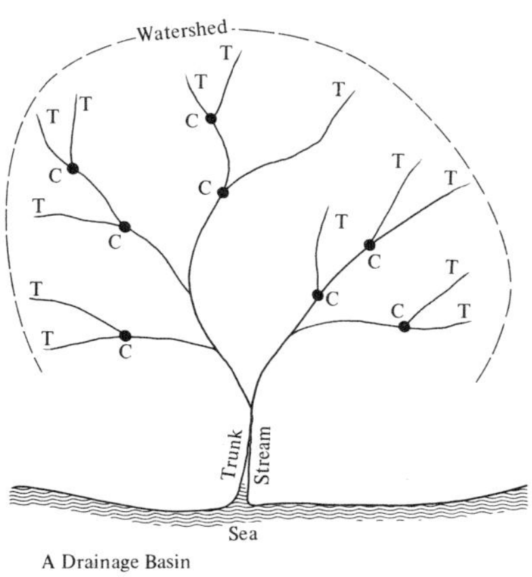

Figure 62

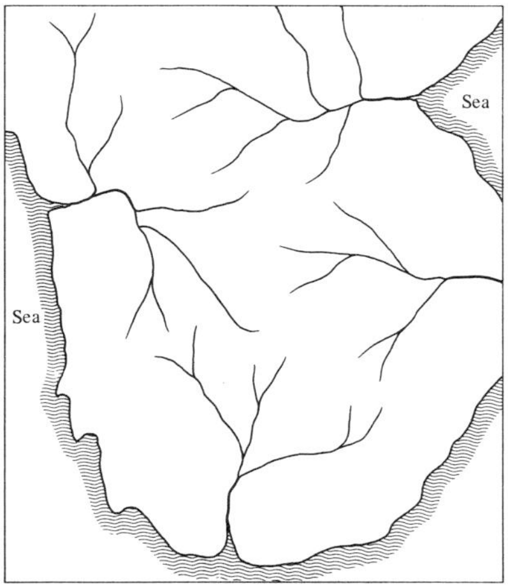

Figure 63

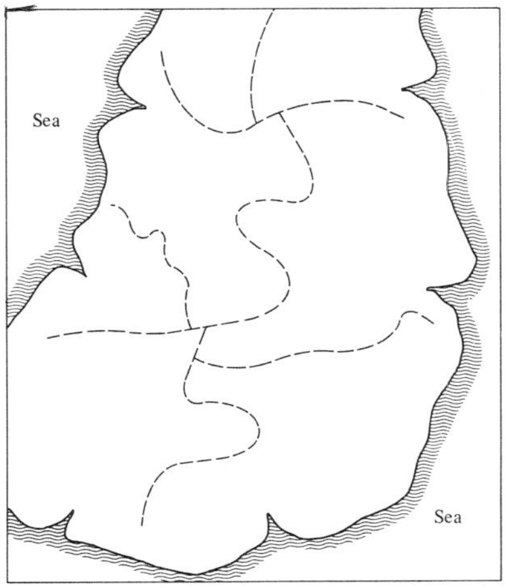

Figure 64

EXERCISE 2: Orders of streams in a river network

The various streams in a river network can be classified roughly according to the amount of water which flows along them, as shown in Figure 65. The smallest streams of all are likely to be the ones which have no tributaries. We call these *first order streams*. They are labelled with a figure 1 in Figure 65. Most of them start near the watershed. When two of these streams join together at a confluence, their flows of water are added together to produce a larger stream. This is called a *second order stream* (1st plus 1st = 2nd). If another 1st order stream flows into this one, a 3rd order stream is produced (2nd plus 1st = 3rd).

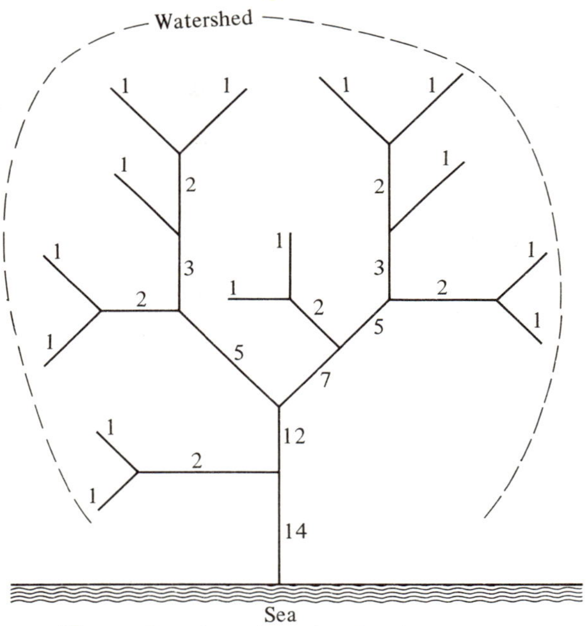

The numbers show the order of each stream

Figure 65

Study Figure 65 and you will see that the order of any stream is the sum of the orders of the two streams which join to form it. Near the sea the river reaches the 14th order because of the confluence of a 12th and a 2nd order stream.

(i) How many *first* order streams provide water for this 14th order stream?

(ii) Make a tracing of Figure 66. Then label each stream with the number of its order.

(iii) Copy Figure 67. In this diagram the trunk stream belongs to the 10th order. Draw a river network which matches this.

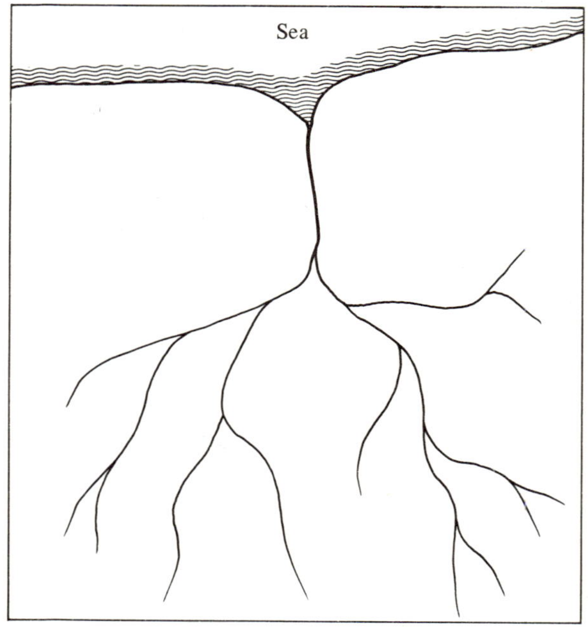

Figure 66

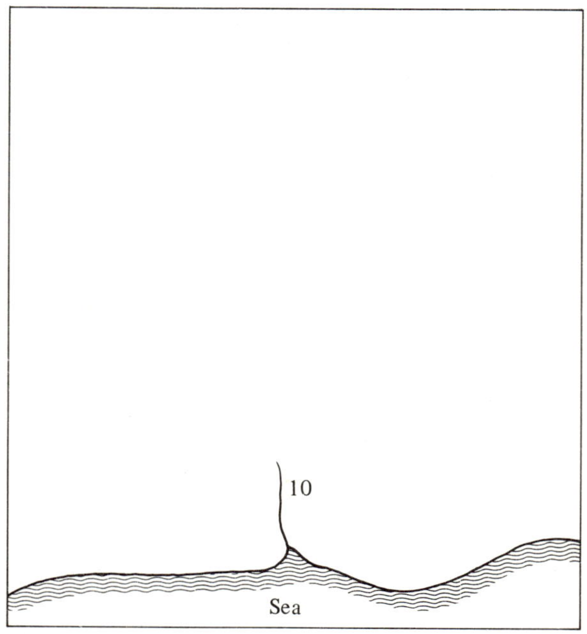

Figure 67

This is stream ordering according to Shreve. Strahlers method is favoured i.e 1+1 = 2 +1 = 2
2+2 = 3 +2 an 1 = 3

EXERCISE 3: The road network of a town

The flow of water to the sea in a stream network is something like the flow of traffic into a city centre when people are travelling to work in the morning. Most towns and cities have branching road networks leading to the centre, like river networks leading to the sea.

We can classify (order) a town's roads in the same way as the streams of a river network. The main difference is that towns have ring roads which join the different branching networks together.

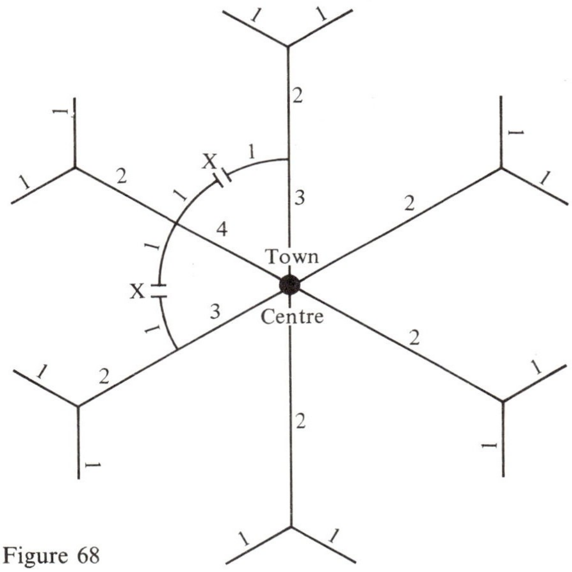

Figure 68

Figure 68 shows the orders of the roads of a town. You will see that the ring road has been divided at the points marked X. Each point X is the same distance from the town centre by each of the two shortest routes. We are supposing that no traffic passes *through* any point marked X but always flows *away from X*. In Figure 68 we expect the 4th order road to carry most traffic. Of course this may not always be true, but it is a simple way of making a rough estimate.

(i) All 1st order roads are lab·· 69 and 70. Place t··· two diagrams an· of all the other r·

(ii) Then draw two maps showing how you expect the volume of traffic to vary in these two areas during the morning rush hour. Use thicker lines for heavier traffic.

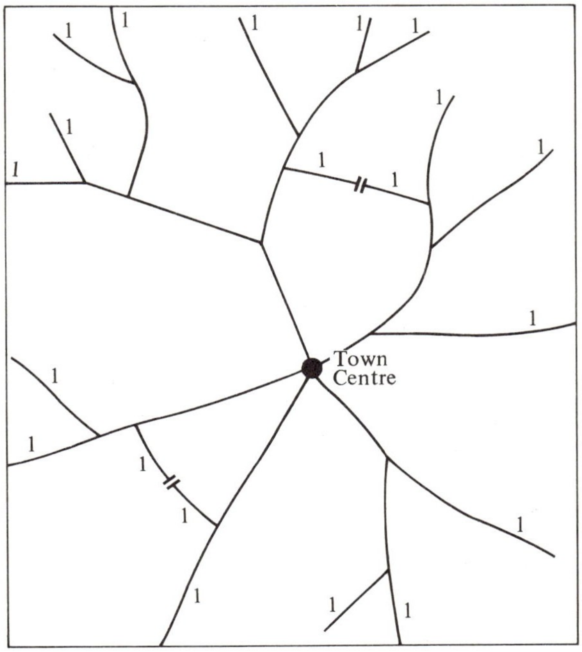

Figure 69

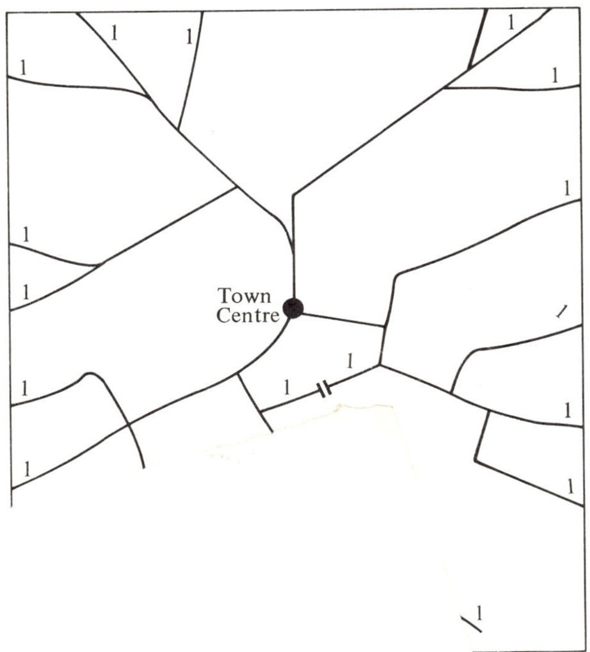

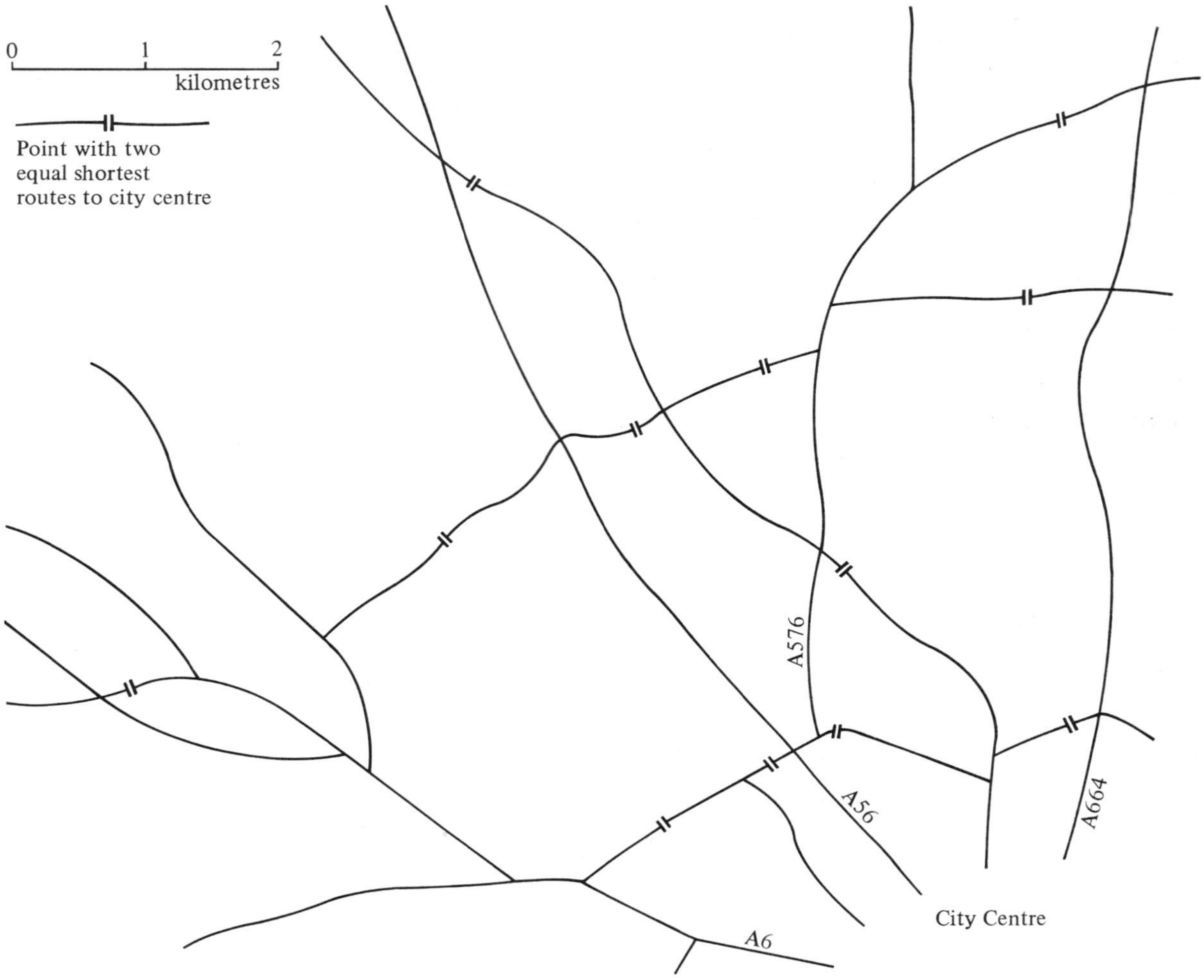

Figure 71

EXERCISE 4: Estimating traffic in north Manchester

Figure 71 shows the pattern of major roads in the suburban areas north of Manchester city centre.

Use the road ordering system to make an estimate of the volumes of traffic which are likely to be carried by the A6, A56 and A664 roads in the morning rush hour.

Give some reasons why your estimate may not be correct.

Unit II: How networks control space

EXERCISE 1: Port hinterlands

Transport networks are developed from ports so that they can attract exports from their hinterlands and supply their hinterlands with imports.

Port hinterlands vary a great deal in size. Large ports like London and Liverpool serve very large areas indeed, much larger than, for example, small Cornish ports. However, for the purpose of this exercise we are defining the hinterland of a port as the area which lies nearer to that port than to any other.

Figure 72 shows how an island may be divided up into the hinterlands of two ports, X and Y. The hinterland of Port X is all nearer to Port X than to Port Y, and the hinterland of Port Y is all nearer to Port Y than to Port X. Two hexagons are equal distances from the two ports, so they have not been included in the hinterland of either. The transport routes linking the ports and their hinterlands pass through the *centres* of the hexagons. Distances from each port are calculated by adding the distances between the centres of neighbouring hexagons. Each hexagon is labelled with its distance (in hexagons) from both ports.

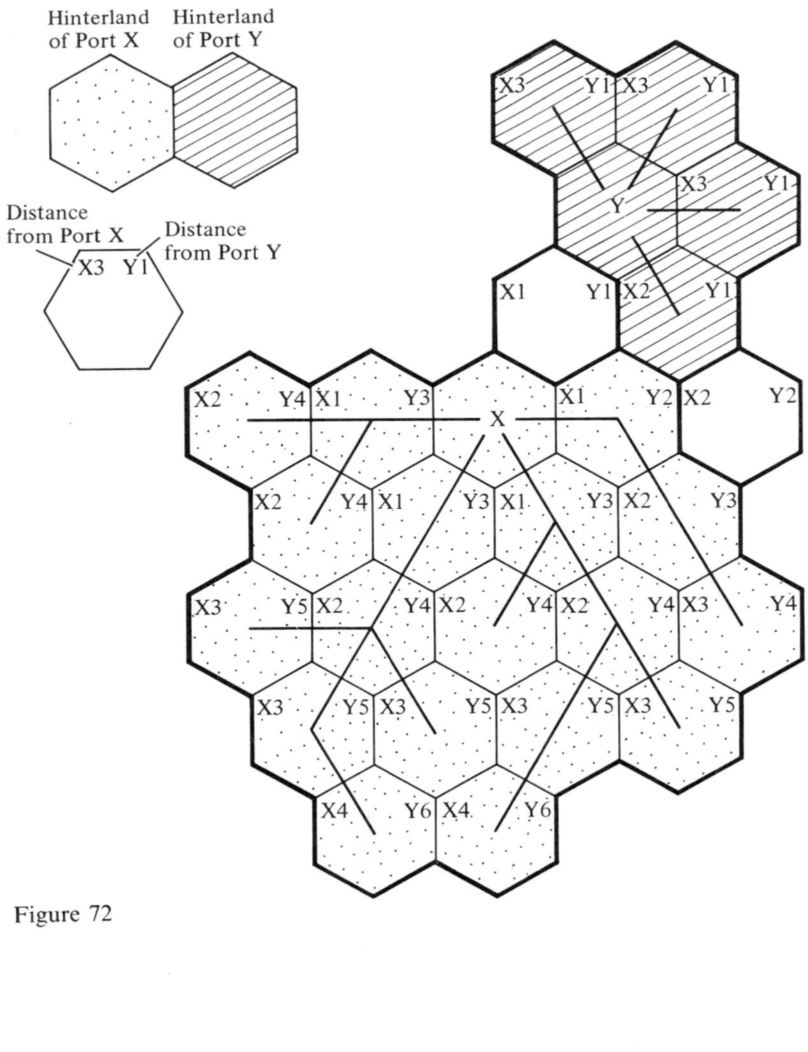

Figure 72

Figure 73 shows a different island with two ports (P and Q). Place a piece of tracing paper over this diagram. Trace its outline. Then divide the island into two hinterlands using the same method as in Figure 72. Colour the hinterlands in different colours, and draw in a suitable transport network.

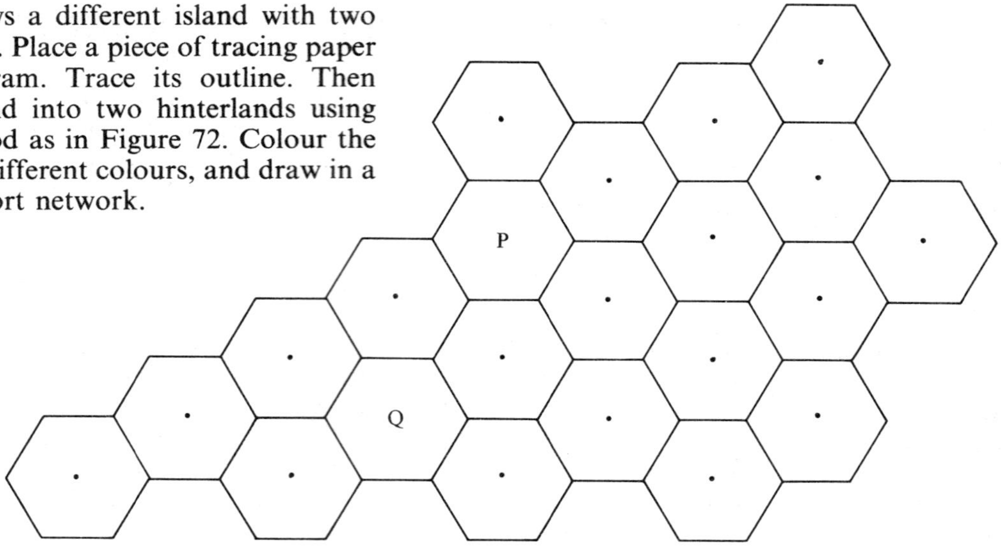

Figure 73

EXERCISE 2: Port competition game (1)

Ports expand or decline as changes take place in the transport networks linking them with their hinterlands. Ports compete with one another for trade. Here is a game about this competition.

Play the game as follows.

(i) Each player takes one of the two ports.

(ii The object of the game is to build a transport network which controls as much as possible of this island. A player controls all the hexagons which are nearer to his own transport routes than to his opponent's.

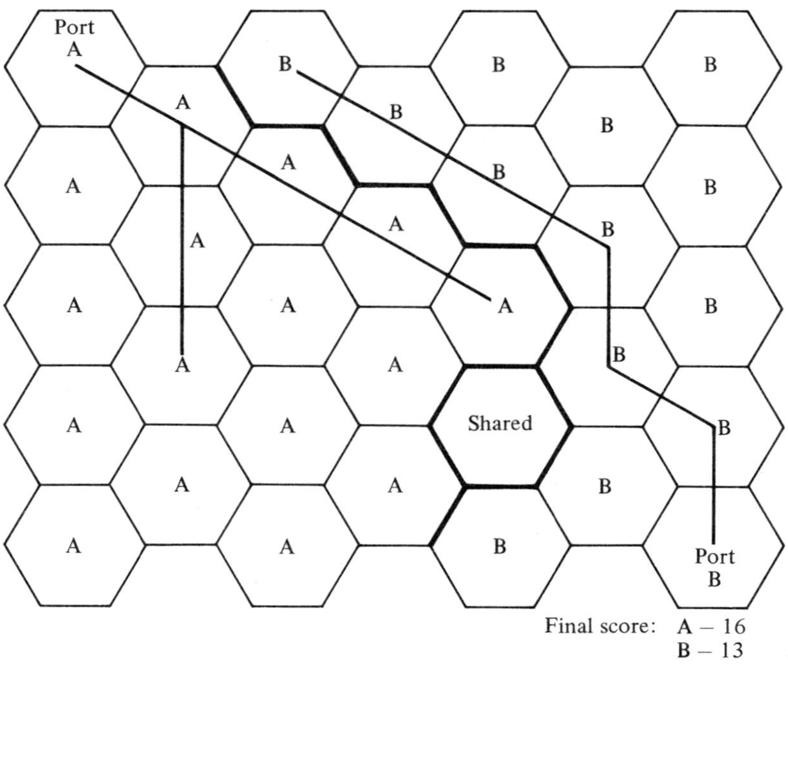

Figure 74

(iii) Each player is allowed to build six transport links from his port. A link is the distance from the centre of one hexagon to the centre of a neighbouring hexagon. All links built by a player must be connected to his port. Neither player may build a link into a hexagon already occupied by his opponent's network.
(iv) The winner of the toss builds the first link. Then the players build links alternately until they each have six.
(v) Figure 74 shows a completed game. In scoring: (i) count the six hexagons occupied by your network, plus all hexagons which are nearer to your network than to your opponent's; (ii) do not count your port, or shared hexagons which are equidistant from each network.

Use a blank diagram to play the game.

EXERCISE 3: Port competition game (2)

Try the same game as in Exercise 2 using a copy of Figure 75 in which the island has an irregular shape. This time the best network is not so easy to find.

Figure 76 is also an island with an irregular shape. This time the players themselves choose where they will locate their ports before starting to build their networks. The winner of the toss has first choice of location for his port. The loser of the toss must also locate his port before any links are built.

In both of these cases each player builds six links.

To play this game use a blank diagram.*

EXERCISE 4: Port competition game (3)

In this game the island is much larger. Three or four players could each choose a port location, or two players could choose two port locations each. In the latter case, the winner of the toss chooses the first port and the rest are chosen alternately.

The methods of play and scoring are exactly as before with the following exceptions:
(i) Each player may build eight links from each of his ports.
(ii) If a player builds a link into a hexagon marked X he may build two links instead of one at his next turn. But he must not build more than eight links altogether from any single port. It may therefore happen that a player may miss a turn towards the end of a game. But he will of course have had the advantage of building his links earlier.

Use a blank diagram to play the game.*

* Figures 75 and 76 are separately reproduced on Card 7. See page 2.

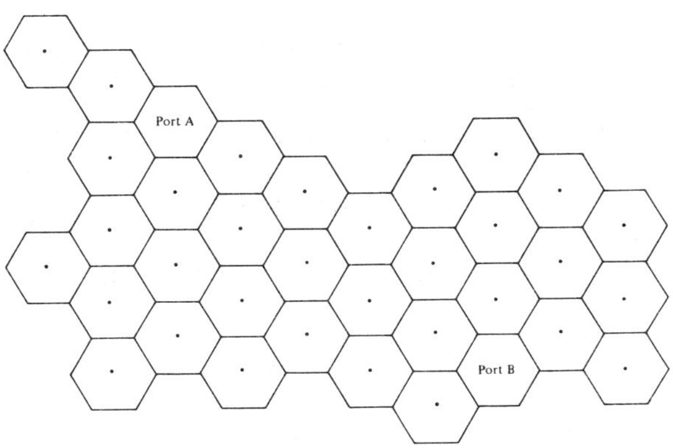

Figure 75

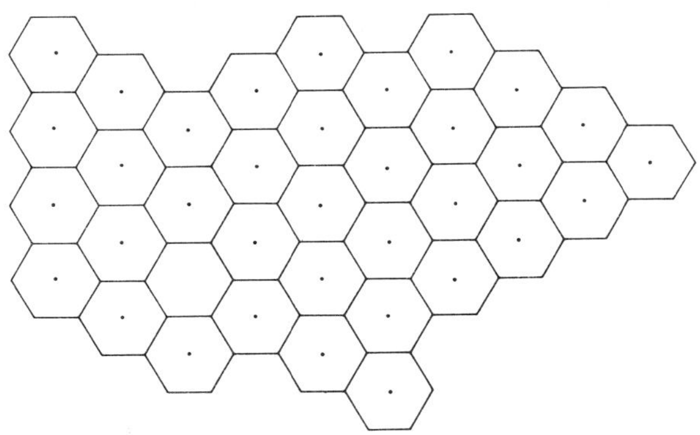

Figure 76

Unit III: Network simulation

EXERCISE 1: The gravity model

The main purpose of a motorway network is to link together the main cities and conurbations in a country, so that people and goods are able to move freely between them. Motorways are particularly useful if they link cities between which a large amount of traffic moves. In planning a motorway network therefore, we must try to work out how much traffic flows from one city to another.

One fairly simple way of doing this is to use the 'gravity model'. By this model we measure the 'interaction' between two cities by multiplying together their population totals and dividing this figure by the square of the distance between them. We would therefore expect a lot of traffic to flow between two big cities and also between two cities which are situated close together. If two big cities are very close together we would expect very large amounts of traffic to flow from one to the other.

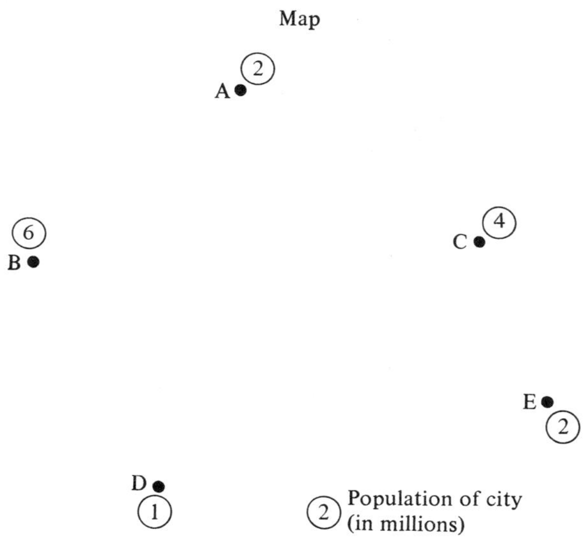

				A
			B	33
		C	24	12
	D	4	17	2
E	2	50	8	3

Interaction matrix

Figure 77

Figure 77 shows five large cities in an imaginary country. We are going to try to design a motorway network to serve their needs. The gravity model has been used to calculate the 'interaction' between each pair of cities, and the results are given in the interaction matrix. Let us first make sure that we understand this matrix.

(i) Between which two cities would you expect the greatest amount of traffic to flow?

(ii) Why do we expect more traffic to flow between C and E than between A and B, although the combined population totals of C and E are smaller than those of A and B?

(iii) Make a copy of Figure 77. Then draw all possible straight lines joining A, B, C, D and E with one another. Using the interaction matrix, label each of these lines with its level of interaction.

EXERCISE 2: Simulating a motorway network (1)

We shall now design a motorway network to serve the cities shown in the previous exercise.

We begin by looking back at the interaction matrix in Figure 77 and inserting motorways along the four links which have the highest interaction values. These are: C to E (50), A to B (33), B to C (24) and B to D (17). These are shown in Figure 78.

You will now see that all the cities are linked to the motorway network, but there is no direct link between, for example A and C (interaction 12) or B and E (interaction 8). Perhaps

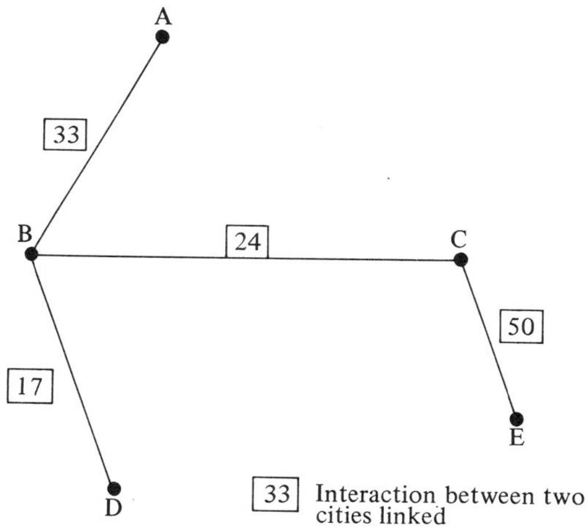

Figure 78

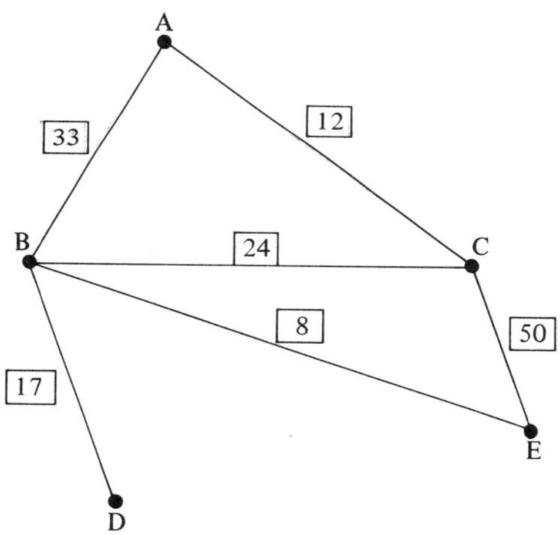

Figure 79

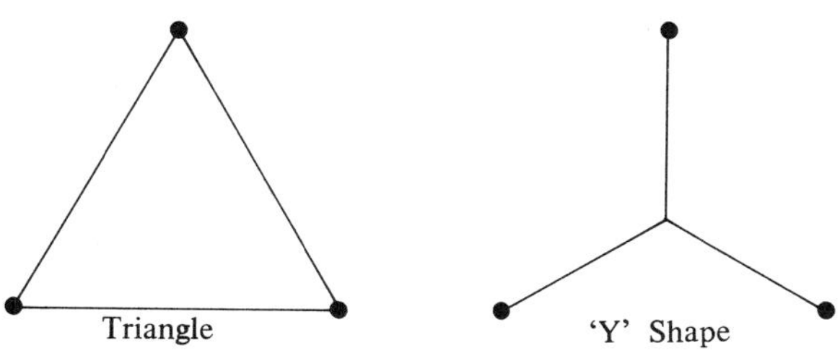

Figure 80

we had better add these links to the network, as shown in Figure 79. There seems no point in adding any more links because the rest of the interaction values are very small indeed.

Figure 79 therefore represents a fairly complete set of motorways for this area. However, because motorways are so expensive to build, perhaps we should try to economize a little, by reducing the total length of motorway. This can be done by changing triangles to Y-shapes, as in Figure 80. A Y-shape has a much shorter length of motorway than a triangle, and is not much less efficient. But the Y-shape should be designed so that the link with the highest interaction level is kept fairly straight.

Figure 81 is the result of changing the triangles in Figure 79 into Y-shapes. You will notice that links AB and BD, with high interaction levels, have been kept fairly straight.

Measure the total length of motorway:
(i) in Figure 79;
(ii) in Figure 81.

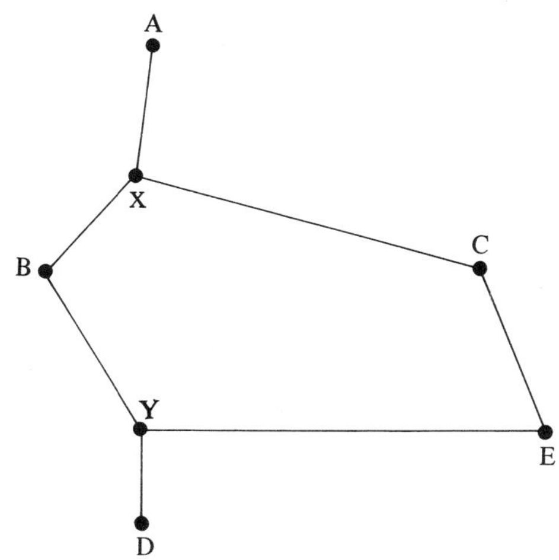

Figure 81

EXERCISE 3: Simulating a motorway network (2)

Now that we have completed our motorway network, let us try to estimate the amount of traffic which is likely to flow along its various links. This can be done by finding the shortest route between each pair of cities and noting the interaction level for each route.

(i) Calculate the total interaction along each motorway link by adding together the interaction values for the inter-city routes which pass along each link. For example, the following routes pass along link AX: A to B (interaction 33), A to C (interaction 12), A to D (interaction 2) and A to E (interaction 3). This gives a total interaction of 50 along link AX. The table below gives the shortest routes which pass along each link.

Link	*Routes*
BX	A to B, A to D, B to C
BY	A to D, B to D, B to E
CX	A to C, B to C, A to E
DY	A to D, B to D, C to D, E to D
YE	D to E, B to E, C to D
CE	C to E, C to D, A to E

(ii) Which three links are likely to be particularly busy?

(iii) Which link is likely to carry least traffic?

(iv) Which of junctions X and Y is likely to be busier?

(v) Draw a diagram like the one below (Figure 82) and show the probable traffic flows along the various links of the network by drawing lines along each link according to the following scheme. Link AX has been completed as an example.

Interaction value	*Number of lines*
11 to 20	2
21 to 30	3
31 to 40	4
41 to 50	5
51 to 60	6

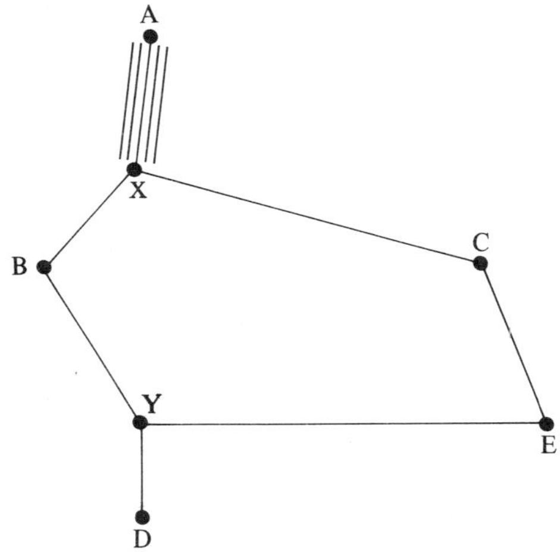

	D	C	B	A
B				33
C			24	12
D		4	17	2
E	2	50	8	3

Interaction matrix

Figure 82

EXERCISE 4: The English motorways (1)

In simulating the motorway network of England we are taking into account levels of interaction on the basis of the gravity model between the great conurbations. We shall include Greater London, the six metropolitan counties created in 1974 (West Midlands, Greater Manchester, West Yorkshire, Merseyside, South Yorkshire, and Tyne and Wear), and the county of Avon, which includes Bristol, the largest city in England outside the metropolitan counties. The central cities of these counties are shown on the map in Figure 83. Their calculated interaction levels with one another are shown in the matrix. These eight counties contain over 64 per cent of England's population, so it seems likely that they have a strong influence on the patterns of road traffic.

Figure 83

Carry out your simulation as follows:

(i) On a sheet of tracing paper trace the dots indicating the locations of the eight cities.

(ii) Draw lines to show all links with interaction values of 5 or over.

(iii) Try to improve the network by replacing triangles with Y-shapes, by combining two parallel links into one motorway, etc.

(iv) Draw your final network on a map of England and Wales. Compare it with the ones produced by your friends.

	Tyne and Wear	South Yorkshire	Merseyside	West Yorkshire	Greater Manchester	West Midlands	Greater London
West Midlands							17
Greater Manchester						12	6
West Yorkshire					35	5	4
Merseyside				6	36	6	3
South Yorkshire			4	25	25	6	4
Tyne and Wear		1	1	3	2	1	1
Avon	0	0	1	1	1	3	5

Interaction matrix

EXERCISE 5: The English motorways (2)

Your simulated network probably looks something like Figure 84.

(i) State three great differences between the simulated network (Figure 84) and the actual network (Figure 85).

(ii) According to the interaction matrix in Figure 83, which is the most obvious link which is missing from the motorway network of England?

(iii) Try to give a special reason for the existence of the following motorways:
(a) M55; (b) M50; (c) M2; (d) M62 east of Leeds; (e) M5 south of Bristol; (f) M6 north of Manchester.

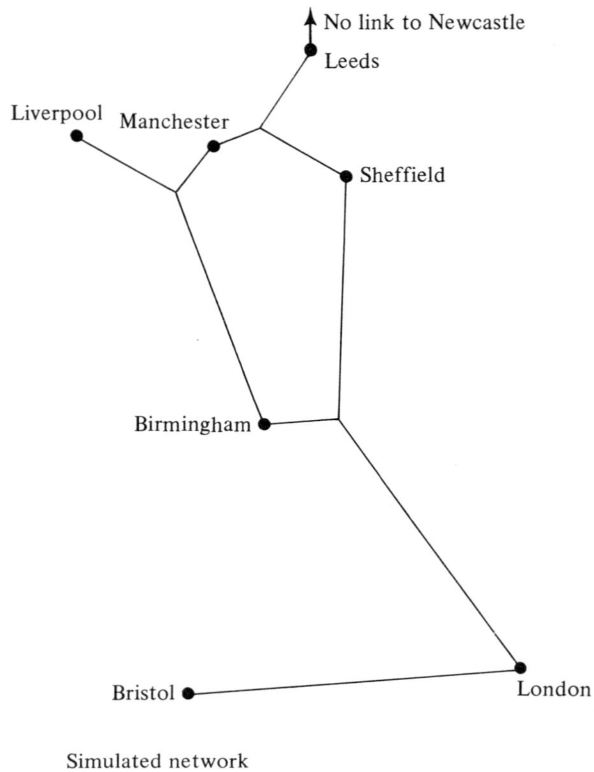

Figure 84

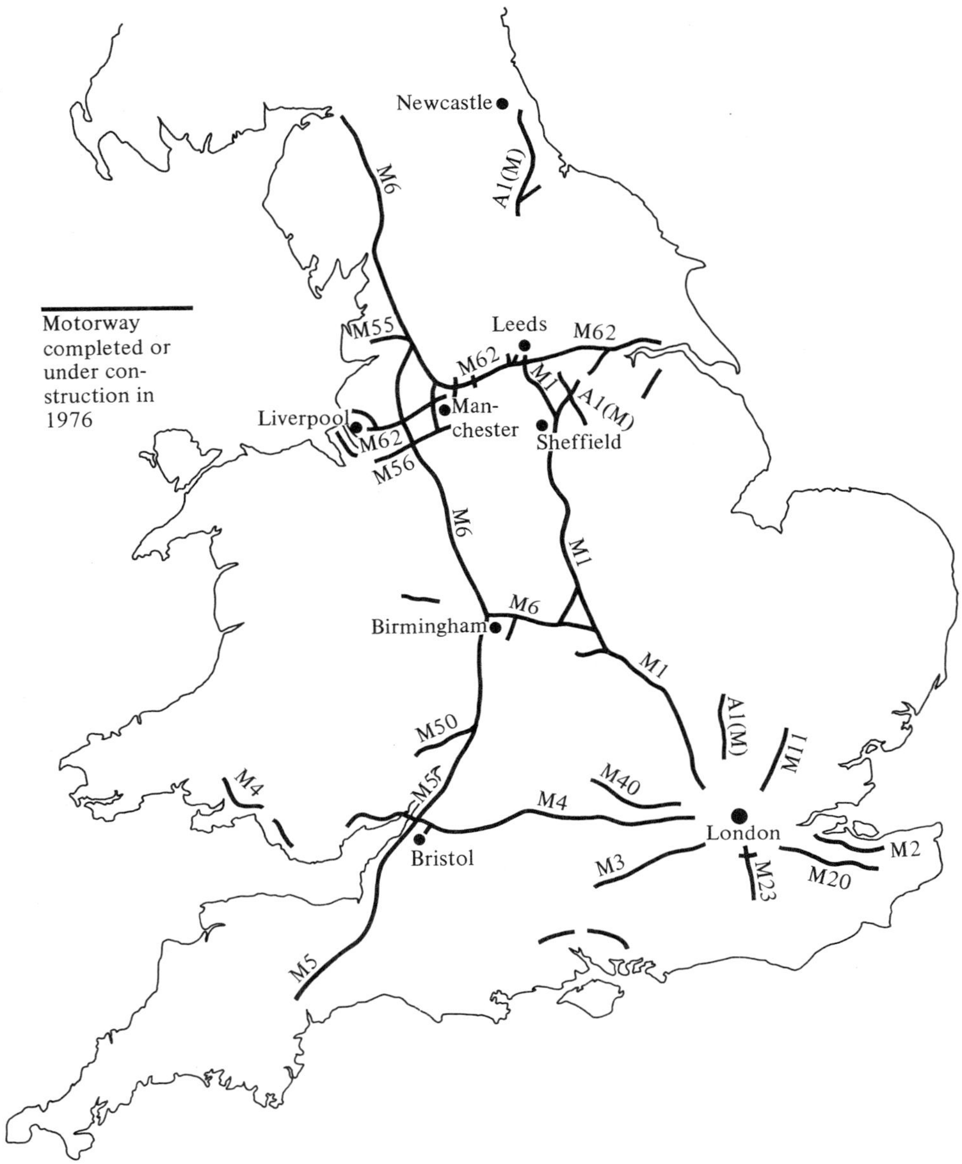

Actual network (England and Wales)

Figure 85

EXERCISE 6: The English motorways (3)

Figure 86 shows the lengths of the main motorway links between the large conurbations of England.

(i) Calculate the distance by motorway:
(a) from Birmingham to London; (b) from Manchester to Sheffield; (c) from Bristol to Liverpool.

(ii) Calculate the length of the shortest possible journey by motorway:
(a) from Leeds to Birmingham; (b) from Birmingham to Sheffield; (c) from Bristol to Leeds.

(iii) Which of the stretches of motorway listed below is likely to carry the greatest amount of traffic? (Work out how many of the inter-city journeys are likely to pass along each motorway link. Then, using the interaction matrix, add up all their interaction values.)

The M6 between Birmingham and the M62;
the M1 between Sheffield and the M69;
the M5 between Bristol and Birmingham.

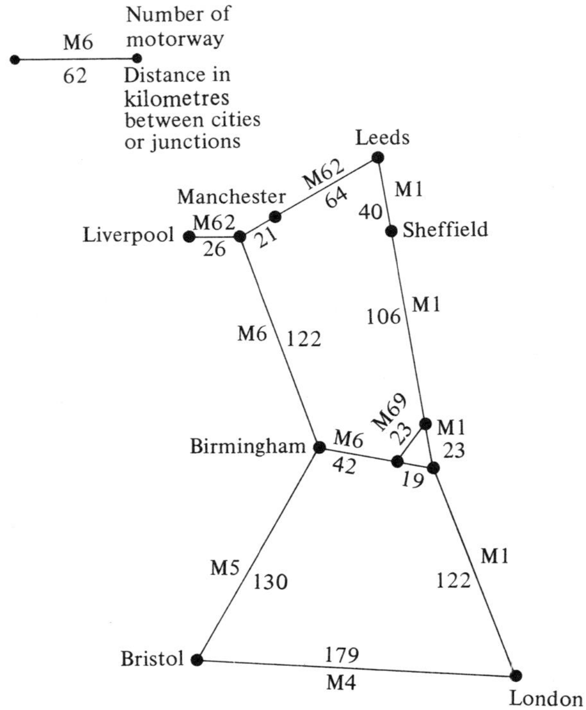

	Newcastle	Sheffield	Liverpool	Leeds	Manchester	Birmingham	London
Birmingham							17
Manchester						12	6
Leeds					35	5	4
Liverpool				6	36	6	3
Sheffield			4	25	25	6	4
Newcastle		1	1	3	2	1	1
Bristol	0	0	1	1	1	3	5

Interaction matrix

Figure 86

Section D
Areas in Geography

Unit I: The shape of areas

EXERCISE 1: Sweden

All countries have different shapes. Some are almost circular; others are very long and narrow. Some are divided into separate parts through having islands offshore.

(i) Search through an atlas to find examples of:
  (a) almost circular countries;
  (b) long, narrow countries;
  (c) divided countries.

Figure 87

In theory the best shape for a country is a circle with the capital city in the centre. This would mean that all parts of the country would be as near together and as near the capital city as possible.

Figure 87 shows the shape of Sweden and its three largest cities. Figure 88 is a circle of the same area as Sweden, with Stockholm, the capital, at the centre. Figure 89 is a combination of Figures 87 and 88. You can see at a glance that Sweden's shape is not very efficient.

Figure 88

Figure 89

Figure 90

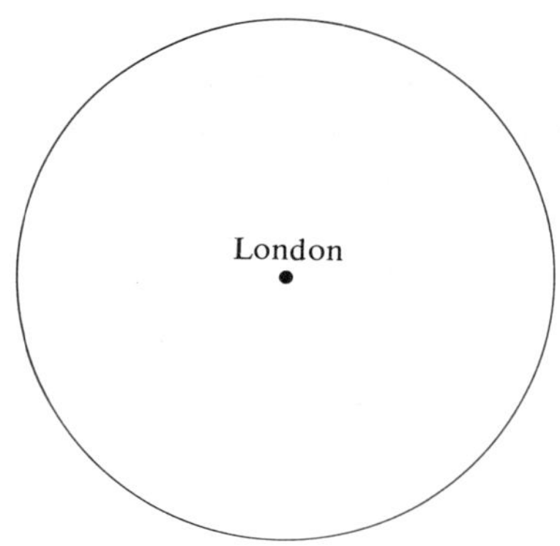

Figure 91

(ii) Draw a sketch map to show the areas on Figure 89 which are further away from Stockholm than they would be if Sweden were circular in shape. These are likely to be the areas where Stockholm's influence is weakest.

(iii) On Figure 89, using the scale, measure: (a) the distance from Stockholm to the furthest point of Sweden (X); (b) the radius of the circle. Then divide the first figure by the second. This will give a measure of the efficiency of Sweden's shape. The larger this figure the *less* efficient the shape.

EXERCISE 2: The United Kingdom

Here is a set of diagrams for the United Kingdom similar to the ones for Sweden in the previous exercise. You will notice that the United Kingdom has a large number of offshore islands and that it also includes Northern Ireland. The cities labelled in addition to London are those with at least half a million inhabitants in 1974.

(i) Name some of the areas on Figure 92 which are further away from London than they would be if the United Kingdom were circular with London at the centre.

(ii) Describe some of the *political* problems which have arisen in these areas.

(iii) Describe some of the *economic* problems which have arisen in these areas.

(iv) Using the method explained in the previous exercise, calculate the shape efficiency of the United Kingdom.

(v) How does this compare with Sweden?

(vi) Which of the cities shown in Figure 90 is located so as to give the best shape efficiency for the United Kingdom?

(vii) Why then do you think that London is the capital?

Figure 92

EXERCISE 3: The U.S.S.R.

The idea of the efficiency of the shape of an entire country might be a bit misleading. Many countries have large areas which are almost uninhabited, so that it does not matter very much if they are remote from the capital city. We should also think about the shape of the relatively densely populated part of the country, sometimes called the 'effective national territory'.

(i) If a circle equal in area to the U.S.S.R. is centred on Moscow, which of the large cities lie outside this circle?
(ii) On which city would you centre this circle so as to include all the relatively densely populated areas?
(iii) Calculate the shape efficiency of the U.S.S.R. with Moscow as capital.
(iv) In what ways is Moscow 'central' within the U.S.S.R.?

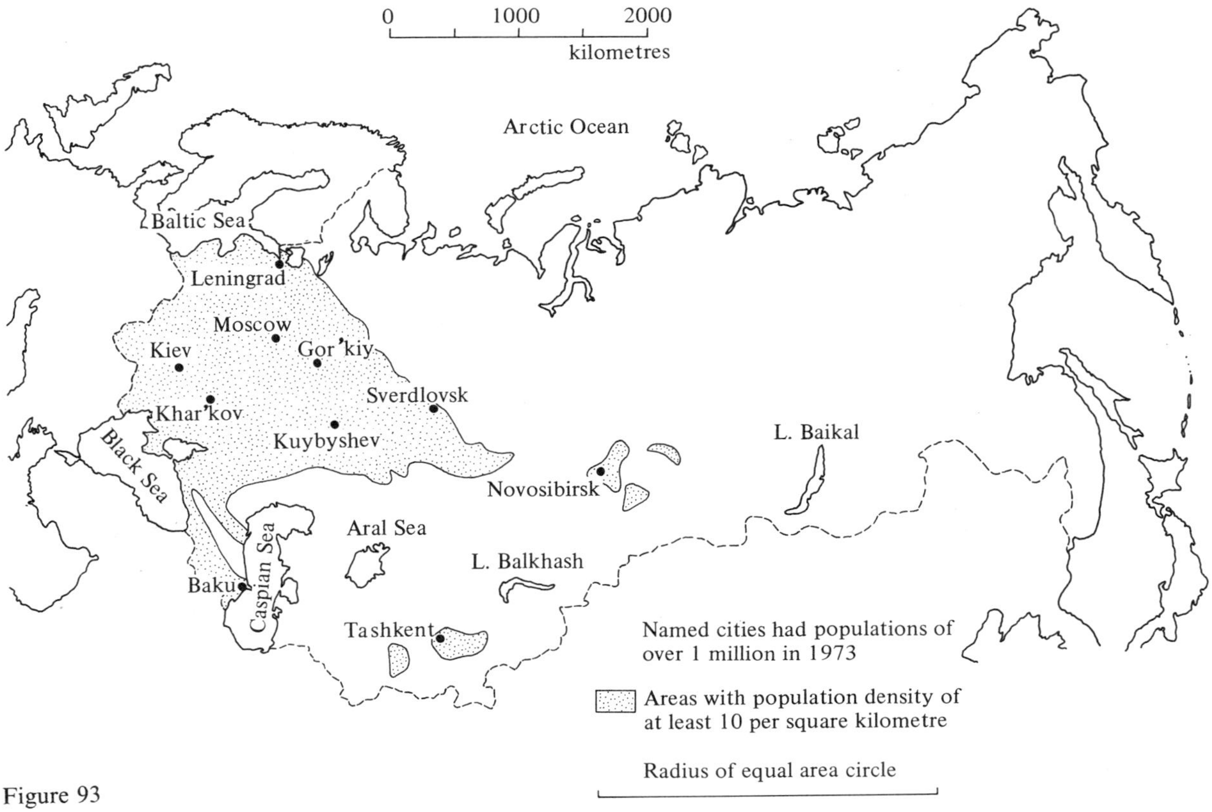

Figure 93

EXERCISE 4: Perception of shape (1)

Very often the shape of a country on a globe or an atlas map is quite different from what its inhabitants 'feel' that it is. Then again, people living in different parts of a country will visualize that it has quite different shapes. It is often said for example that people who live in London tend to regard everywhere north of Watford as a vaguely defined area best called 'the North'. Lancastrians and Yorkshiremen may be similarly somewhat vague about the area to the south of the Trent. We may be able to make maps of countries as people 'feel' that they are. Figure 94 is an 'atlas-type' map of mainland Scotland with eleven towns (two of which are in England) scattered fairly evenly. Its scale is in kilometres, a measure of distance.

Figure 95 is a 'mental map' such as a frequent train traveller living in Edinburgh might 'see' the shape of Scotland. Its scale is in hours of train travel time from Edinburgh. It is drawn as follows.

(i) First the dotted lines are drawn from Edinburgh in the directions of the other towns.

(ii) Then the travel time to each place is marked off along each dotted line.

(iii) Finally the coastline is drawn in as well as possible between the fixed points of the towns. The coast will obviously be considerably distorted.

Explain briefly how and why the shape of the map in Figure 95 differs from that in Figure 94.

Figure 94

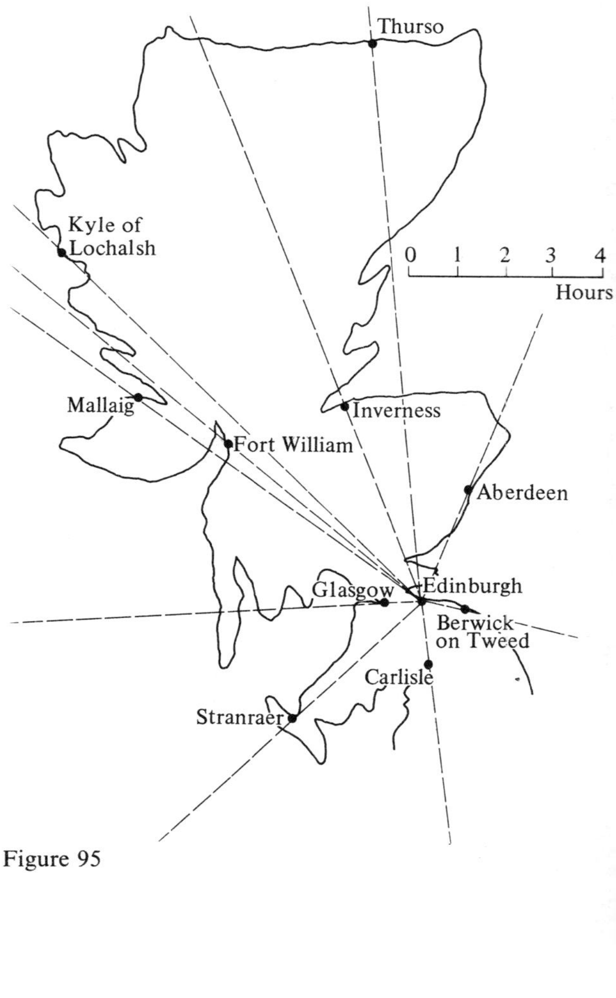

Figure 95

EXERCISE 5: Perception of shape (2)

Let us now try to find out whether Scotland would seem to have a different shape to a frequent train traveller living in Inverness. The dotted lines in Figure 96 show the directions of all the other towns from Inverness. Make a tracing of this diagram and mark off the positions of these towns using the following travel times by train from Inverness in 1974–75.

	Hour	*Minutes*
Aberdeen	2	30
Berwick on Tweed	5	29
Carlisle	5	57
Edinburgh	4	24
Fort William	9	10
Glasgow	4	0
Kyle of Lochalsh	2	23
Mallaig	11	9
Stranraer	7	42
Thurso	4	39

Now draw your 'coastline' trying as far as possible to show the various headlands and inlets.

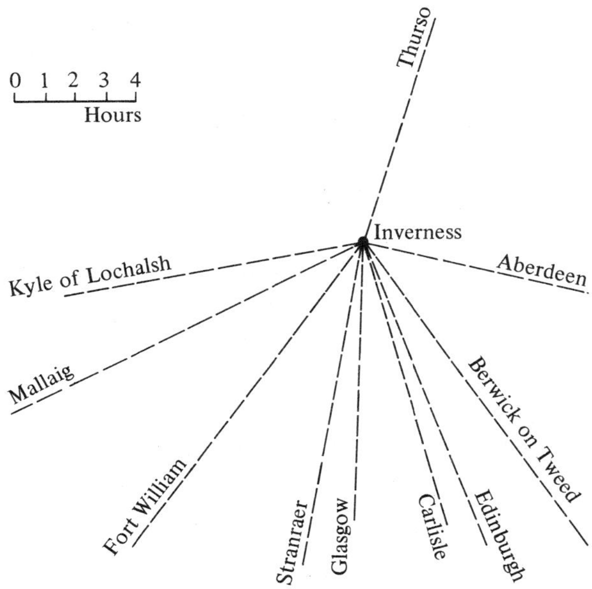

Figure 96

Explain briefly how and why your finished map differs from:

(i) Figure 94;
(ii) Figure 95.

Now get a railway timetable and try to construct a similar map based on travel times from your own town to a number of places on the coast.

Unit II: How to classify areas

EXERCISE 1: Classification by size

All areas are different from one another, but some of them are sufficiently alike in some respects for us to be able to group them together and thus build up larger 'regions'. Each of these regions will be composed of areas which are broadly similar in some ways.

Let us use the countries of the E.E.C. (European Economic Community) as our areas. We are going to divide them into 'types' according to their size.

The sizes of all the countries are shown on the linear scale (Figure 97). Belgium and Luxembourg have been taken as a single area. Perhaps you did not know that France is just about as big as any other two countries added together. It is easy to see that these countries divide up into three size groups.

(i) Which are:
  (a) the large countries; (b) the medium-sized countries; (c) the small countries?

(ii) On an outline map of the E.E.C. colour the three kinds of countries in different colours and give a key. Label each country by name, and use an atlas to allocate the islands to their respective countries.

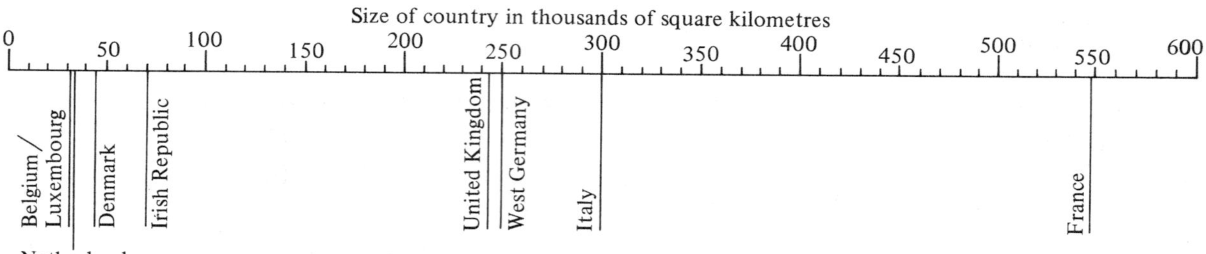

Figure 97

EXERCISE 2: Classification by population total

Let us now group these same countries of the Common Market in a different way. This time we shall use their population totals, which are given in the table below.

Country	*Population (millions)*
Belgium/Luxembourg	10·1
Denmark	5·1
France	52·5
West Germany	62·0
Irish Republic	3·1
Italy	55·4
Netherlands	13·5
United Kingdom	56·0

(i) Draw a scale like Figure 98 and mark the countries on it.

(ii) How many groups of countries are there in this case?

(iii) Refer back to the previous exercise to find the areas of France and West Germany. Which of these two countries has the higher *density* of population?

(iv) Colour an outline map of the E.E.C. to show your groups of countries according to population total. Insert and label a key.

Figure 98

EXERCISE 3: Classification by size and population total

We have now looked at the countries of the E.E.C. in two different ways:
(a) Size, by which we could recognize *three* types;
(b) Population totals, by which we could recognize *two* types.

If we combine these two characteristics we shall find that there are *six* possible types of countries (3 multiplied by 2). These are shown by the six 'boxes' in the table below.

Population		*Size*	
	Small	*Medium*	*Large*
Large	1	2	3
Small	4	5	6

(i) Which countries would you put in: (a) Box 1; (b) Box 2; (c) Box 3; (d) Box 4; (e) Box 5; (f) Box 6?
(ii) Colour an outline map of the E.E.C. and add a key to show these new groups of countries based on *both* size and population.

If we want to group areas according to two factors like this, we cannot use a single-line scale as in the two previous exercises. We have to use *two* single-line scales at right-angles to one another. This is called a *graph*.

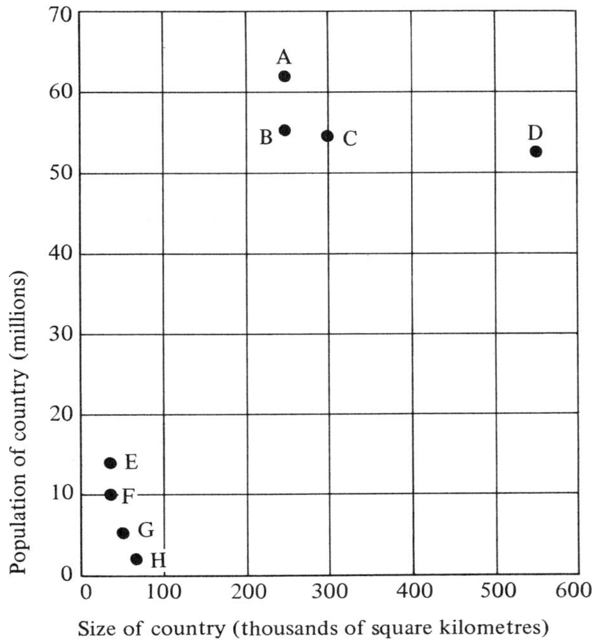

Each lettered dot on the graph on this page represents a Common Market country.
(iii) Which countries are represented by the letters A to H?

You will see that countries in the same box (above) are close together on the graph.

EXERCISE 4: Classification by living standard and population density

Now let us try to group the larger countries of South America according to two of their characteristics. In this case we shall use:
(a) Gross Domestic Product per person. This is really the total value of goods and services produced per person in a year. It is an approximate measure of the standard of living.
(b) Population density per square kilometre. This is the total population divided by the area of the country.

Country	*G.D.P. per person (in dollars)*	*Population density per sq km*
Argentina	1260	9·0
Bolivia	221	4·9
Brazil	513	12·0
Chile	649	14·0
Colombia	474	21·0
Ecuador	386	25·0
Paraguay	373	6·3
Peru	379	12·0
Uruguay	794	17·0
Venezuela	1579	13·0

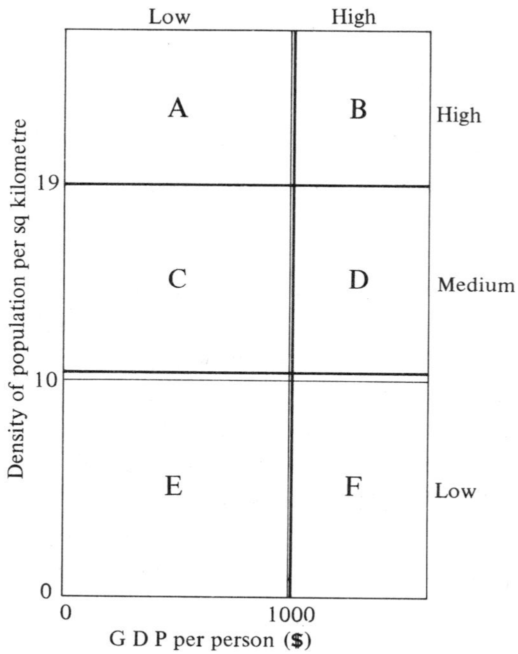

Figure 99

(i) Which countries would you place in each of the classes A, B, C, D, E and F on the graph (Figure 99)?
(ii) Make a tracing of the map (Figure 100) and colour each group of countries differently. Label the name of each country and insert the key.

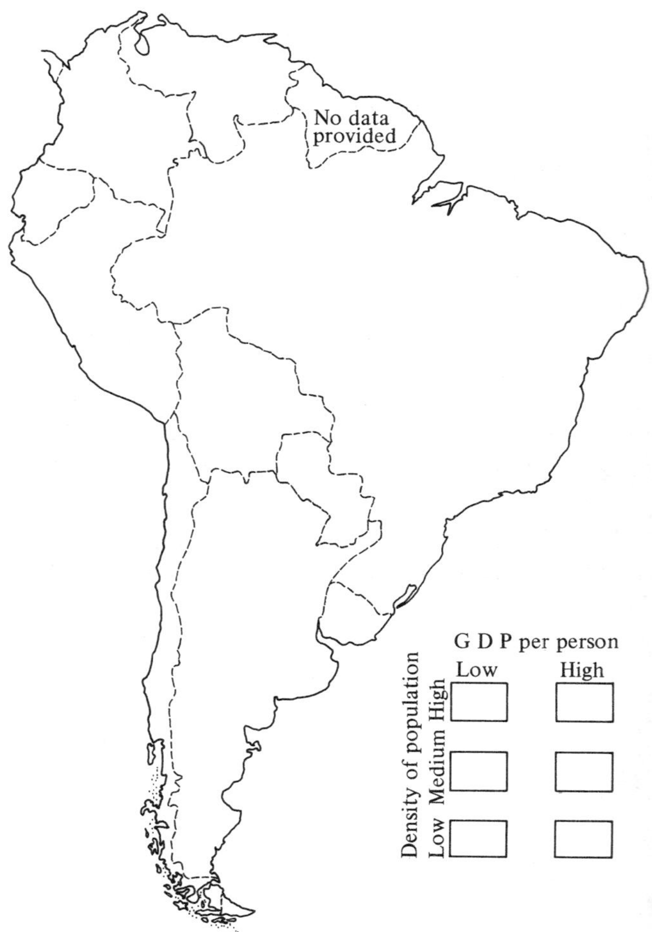

Figure 100

EXERCISE 5: Classification by land-use (1)

When we are grouping areas together to form regions the more factors we use, the better. If we use three factors we shall need three single-line scales. Thus, the rectangular graphs of Exercises 3 and 4 cannot be used. We now need a *triangular graph* like the one in Figure 101. It is useful to be able to use three factors because we can then group our countries according to their land-use pattern, by the following characteristics:

(a) the percentage of the country's area used for arable farming;
(b) the percentage of the country's area used for pasture;
(c) the percentage of the country's area which is uncultivated. Land may be uncultivated because it is mountainous, swampy, forested or occupied by towns, etc.

These three characteristics give us a good general idea of what a country is like.

Let us learn how to use a triangular graph. The U.S.A. has the characteristics shown below:

Percentage of area			
Arable	*Pasture*	*Uncultivated*	*Total*
21	26	53	100

To use a triangular graph it is essential that the three numbers add to 100.

The dotted lines on the graph show how these figures have been plotted.

Make a tracing of the triangular graph and plot the following countries on it.

	Percentage of area		
	Arable	*Pasture*	*Uncultivated*
Denmark	63	8	29
South Africa	6	67	27
Canada	4	2	94

Note that Denmark is nearest to corner X (high percentage arable), South Africa is nearest to corner Y (high percentage pasture), and Canada is nearest to corner Z (high percentage uncultivated). Why does Canada have so large an uncultivated area?

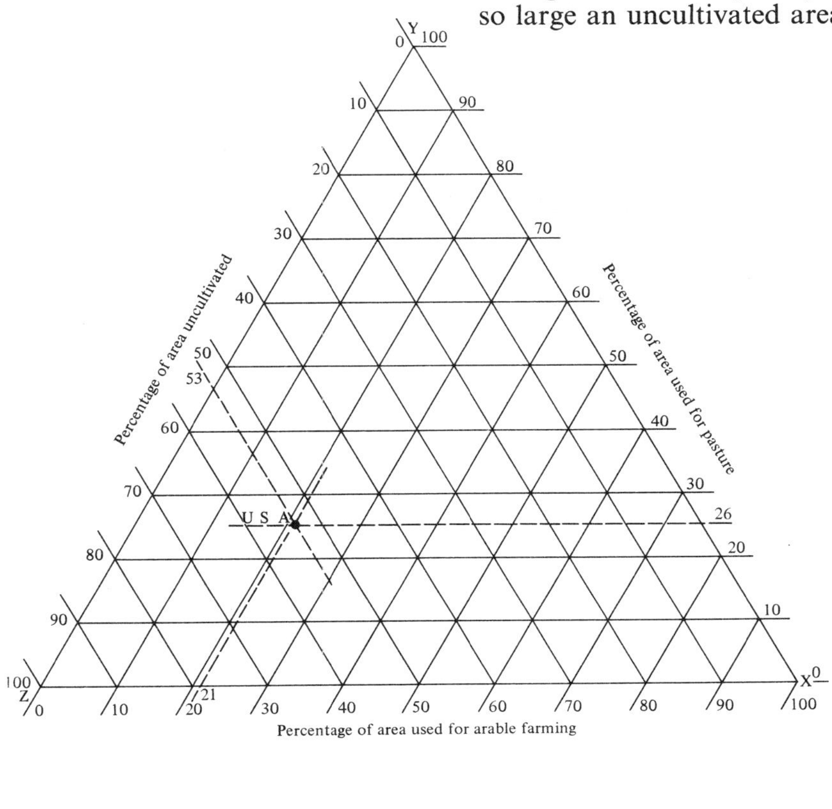

Figure 101

EXERCISE 6: Classification by land-use (2)

We shall now group some of the countries of Europe into regions using the methods explained in the previous exercise. The table below gives the percentages of the area of each country used for arable farming, pasture, and which are uncultivated. The triangular graph (Figure 102) shows seven types of countries (A, P, U, AP, PU, AU and APU).

Allocate each of the countries in the table to one of these seven types. You will find that the Irish Republic lies on a type boundary. Classify it as PU. Write your classification in your exercise book.

	Percentage of area		
	Arable	*Pasture*	*Uncultivated*
Austria	19	26	55
Belg./Lux.	27	24	49
Denmark	63	8	29
Finland	9	1	90
France	34	25	41
West Germany	33	22	45
Iceland	–	23	77
Irish Republic	17	53	30
Italy	42	18	40
Netherlands	25	37	38
Norway	3	1	96
Portugal	49	6	45
Spain	42	23	35
Sweden	7	2	91
Switzerland	10	41	49
United Kingdom	30	48	22

EXERCISE 7: Classification by land-use (3)

(i) Use your classification from Exercise 6 to colour an outline map to show the variations in land-use over these countries of Europe. Insert a key. Draw a thick line round *all* the countries which have the well-balanced APU type of land-use. You will notice that these countries form a solid block. Draw another thick black line round the three countries wich are of the U type (mainly uncultivated) and which again form a solid block.

(ii) Why is there so much uncultivated land:
  (a) in northern Europe;
  (b) in Austria and Switzerland?

(iii) Suggest reasons why only Denmark is of the A type (mainly arable).

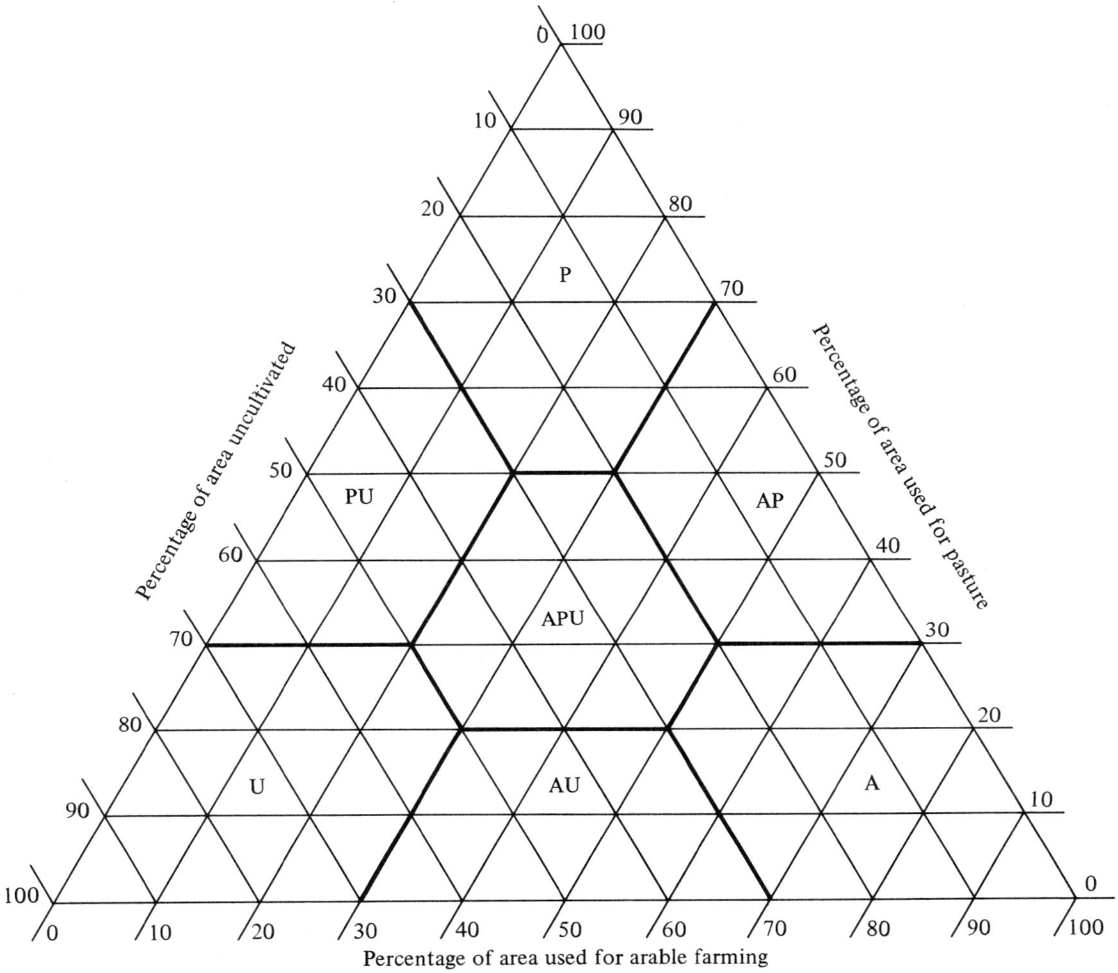

Letter on graph	Description
A	Mainly arable farming
P	Mainly pasture
U	Mainly uncultivated land
AP	Mainly arable farming and pasture
PU	Mainly pasture and uncultivated land
AU	Mainly arable farming and uncultivated land
APU	Well balanced. No particular specialization

Figure 102

Unit III: Developed and developing countries

EXERCISE 1: Employment characteristics

It is sometimes said that countries in which the standard of living is very low tend to have a large percentage of their populations employed in agriculture. These are known as *developing countries*. A sign of a *developed* or *advanced country* is a high percentage of its population working in 'service industry' which includes transport, the retail trade and professions, and in manufacturing industry.

(i) Make a list of the countries given in the table on the right, and mark against each country's name:
(a) the number on the world map (Figure 103) which shows its location;
(b) the group of letters on the triangular graph (Figure 104) which shows its type.

(ii) Which of these countries do you think has the lowest standard of living?

(iii) Which of the countries would you classify as developing countries?

	Percentage of working population employed in:		
	agriculture	*industry*	*services*
Argentina	16	32	52
Australia	8	35	57
Ecuador	59	18	23
India	72	12	16
Indonesia	66	10	24
Libya	42	22	36
Mexico	42	24	34
Morocco	58	14	28
New Zealand	11	36	53
Tanzania	46	21	33
U.S.A.	4	34	62
Zambia	33	24	43

Figure 103

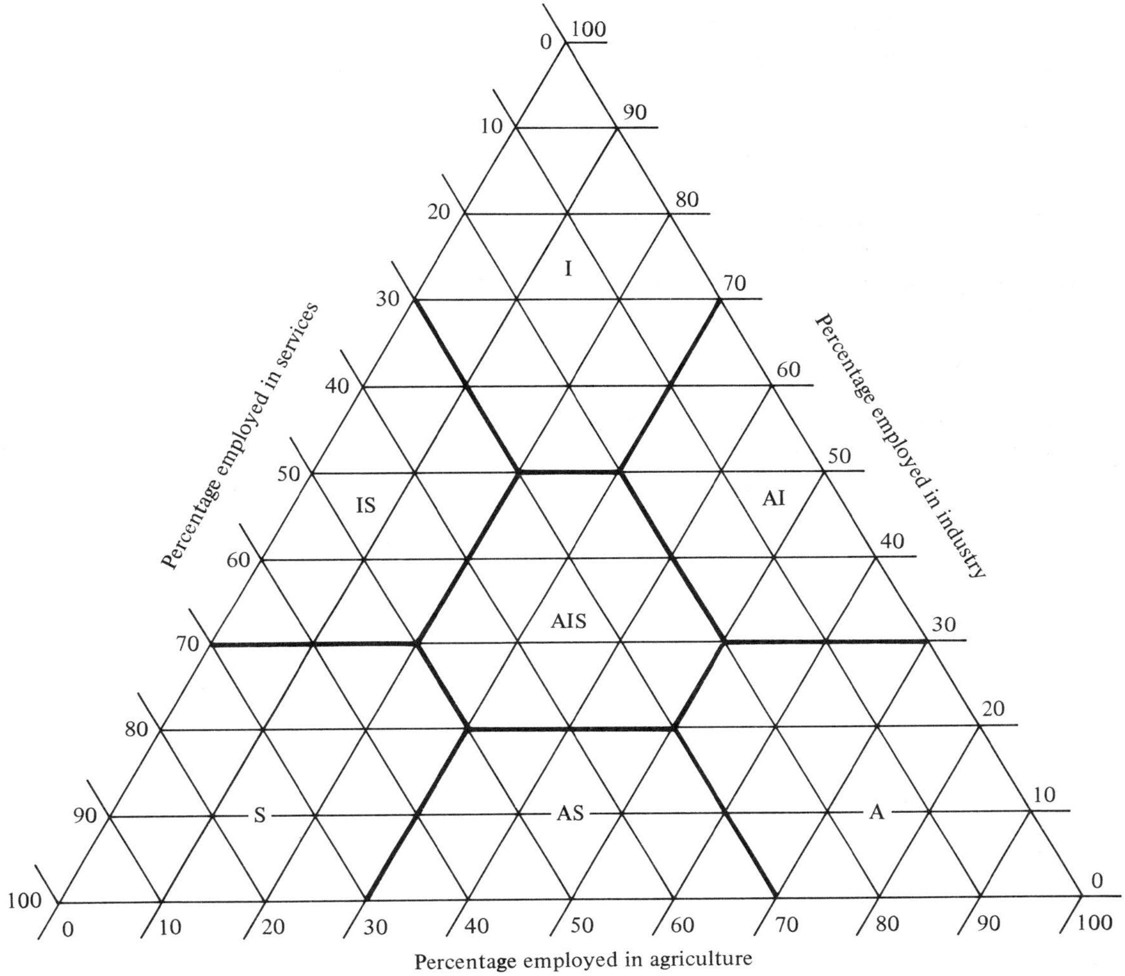
Percentage employed in services
Percentage employed in industry
Percentage employed in agriculture
I
IS
AI
AIS
S
AS
A
0
10
20
30
40
50
60
70
80
90
100

Figure 104

EXERCISE 2: Relations between employment and living standards

As we saw in Unit II, Exercise 4, the standard of living in a country is sometimes measured by the Gross Domestic Product per person. Let us try to find out whether it is true that countries with a large percentage of their populations employed in agriculture tend to have low living standards.

The table in the previous exercise gives the percentage of the population of twelve countries which is employed in agriculture. The table below gives information about living standards in the same twelve countries.

	G.D.P. per person (thousands of dollars)
1. Argentina	1·3
2. Australia	3·7
3. Ecuador	0·4
4. India	0·1
5. Indonesia	0·1
6. Libya	2·3
7. Mexico	0·9
8. Morocco	0·3
9. New Zealand	4·0
10 Tanzania	0·1
11. U.S.A.	6·2
12. Zambia	0·4

(i) Label a sheet of graph paper as in Figure 105, and mark numbered dots to show each country in terms of employment in agriculture and living standards. Then draw a smooth curve through the following dots: 4, 5, 8, 7, 9 and 11. This curve gives a good summary of the general pattern of the dots.

(ii) What do you conclude about the relation between employment in agriculture and living standards?

(iii) Which country has a higher living standard than you would expect? Suggest a reason.

(iv) Now, using information from the table above and from the previous exercise, let us try to find out whether countries with high levels of employment in service industries tend to have high living standards. Label another sheet of graph paper as shown in Figure 106 and mark a numbered dot for each country. Then draw a smooth curve through 4, 3, 7, 2 and 11. What are your conclusions?

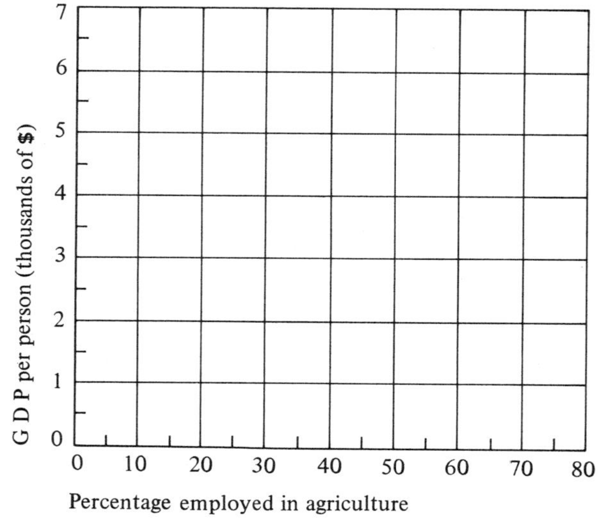

Figure 105

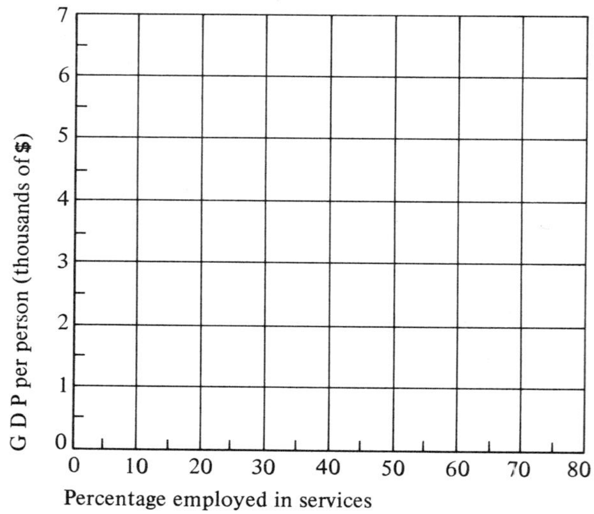

Figure 106